Science and Psychoanalysis. *Volume XIII.*

THE DYNAMICS OF DISSENT

Scientific Proceedings of the
American Academy of Psychoanalysis

Edited by

JULES H. MASSERMAN, M.D.
Professor and Co-Chairman of Psychiatry
Northwestern University
Chicago, Illinois

GRUNE & STRATTON
New York • London • 1968

Preceding volumes of *Science and Psychoanalysis*

Volume I: *Integrative Studies* (1958)
Volume II: *Individual and Familial Dynamics* (1959)
Volume III: *Psychoanalysis and Human Values* (1960)
Volume IV: *Psychoanalysis and Social Process* (1961)
Volume V: *Psychoanalytic Education* (1962)
Volume VI: *Violence and War* (1963)
Volume VII: *Development and Research* (1964)
Volume VIII: *Communication and Community* (1965)
Volume IX: *Adolescence, Dreams, and Training* (1966)
Volume X: *Sexuality of Women* (1966)
Volume XI: *The Ego* (1967)
Volume XII: *Animal and Human* (1968)

Library of Congress Catalog Number 58-8009

381 Park Avenue South
New York, New York 10016

Printed in the United States of America (K-B)

CONTENTS

Part IV. Social Dynamics

PREFACE: THE SEMANTICS OF DISSENT

THIS BOOK is a volume of words; but are these words—to paraphrase Scripture—resounding trash or thinking symbols?

As do symphonic virtuosi, psychiatrists experienced in their art (and Webster's Dictionary defines art as "skill derived from knowledge and practice") acquire an increasing respect for the instrument they employ. Not their cultist jargon, but the deeply meaningful significance of Basic English (or German, or Spanish, or especially French)—languages which retain the eloquence and resonant overtones of aeons of human experience. Indeed, the word *experience* itself means *that which pierces through* to the human "mind," whereas the latter term, taken as process rather than as an abstraction, signifies (a) sentience ("mind what happens!"), (b) memory ("reminded!"), (c) compliance ("mind, i.e., obey, the laws and mores!"), and withal, (d) creativity. Classical mentors recognized that all of these functions were necessary to achieve *mens sanis in corpora sano*—"a healthy mind in a healthy body," whereas in our modern usage, "health" itself is derived from the Anglo-Saxon *hal* or *hol,* comprising the ultimate (Ur) aspirations of man; physical well-being (hale) social alliance (hail friend!) and philosophic/theologic attainments (spiritually *wholesome* or *holy*). To continue in this rich philologic vein, the quasi-mystical adjective "spiritual" has its own illuminating significance: mortals *aspire* (*L. spiritus,* breath) to live with their first *aspiration,* must soon learn *esprit de corps* for social survival, become *inspired* toward technical or artistic contributions, become *exasperated* or *dispirited* if they fail to meet inculcated standards and—whether or not the physical, social, or philosophic Ur-faiths give them interim comfort—by tragic existential decree they inevitably *expire.*

This heuristic excursion into semantics has special reference to the title of this book, since man's rational or irrational rebellions against physical, social, and philosophic duress variously constitute the Dynamics of Dissent. Again derivatively, *dynamics* connotes a generative process (*dynamo*), whereas *dissent* means any process deviant from or contrary to prevailing concepts and goals. Dissenters then expressed themselves in *protests* (preliminary tests of action) and *a-ggressions* or motion toward an objective. But since "aggression" can become destructive, it merits special consideration.

Biodynamically, lower animals have neural nodes aligned throughout their diencephalon and paleocortex which, when stimulated, activate the external mimetics of food-capturing, sexual, or combative action. Nevertheless, despite current neglect of Occam's dictum: *Essentia non sunt multiplicanda praeter necessitatem,* this no more necessitates the postulate of a separate "aggressive instinct" in animal or man than the presence of lateral quadrigeminal bodies bespeak an "instinct for hearing"; on the contrary, as in all CNS functions, the neural facilities are integrated into uniquely holistic receptive—mnemonic—interpretive—responsive patterns. Thus, most studies in our own and other laboratories confirm ethologic observations to the effect that "aggressive" behavior becomes *violent*—i.e., violates the possessions or well-being of others—only when the animal's needfulfilling patterns of conduct are subjected to severe physical or interorganismic deterrents. In human beings such frustrations, which may assume highly complex and symbolic forms, again fall principally into the three ultimate rubrics of concern (Ur-anxieties) enumerated above. In effect, whenever we apprehend major threats to our physical well-being, or to the cultural-political alliances we consider necessary to our social security, or to our cherished philosophic-theologic faiths, we progressively express "dissent," act in "protest" and, if *desperate,* become "violent."

These, then, are the basic concepts and the ranges of human behavior that the contributors to this volume explore: from a neonate's first protest against a newly inimical milieu, through a patient's accouched complaints about his complexes, to the virulent distaste most of us feel toward the violence of war. That each contribution is thoughtful and thorough I can here predict—and the reader will confirm.

Jules H. Masserman, M.D.
Editor

January 1, 1968
Chicago, Illinois

CONTRIBUTORS

ALLEN, JAMES R.: Assistant Professor of Child Psychiatry, University of Oklahoma, School of Medicine, Oklahoma City, Oklahoma.

BAITTLE, BRAHM: Chicago Psychoanalytic Institute, Chicago, Illinois.

BRYT, ALBERT: Assistant Clinical Professor of Psychiatry, New York University, New York, New York.

COHEN, SHELDON B.: Psychoanalyst, Atlanta, Georgia.

COSER, LEWIS A.: Professor of Sociology, Brandeis University, Waltham, Massachusetts.

DARROW, CHARLOTTE: Department of Psychiatry, Wayne State University, Detroit, Michigan.

GALDSTON, IAGO: Fairfield Hills Hospital, Newton, Connecticut.

KENISTON, KENNETH: Associate Professor, Department of Psychiatry, School of Medicine, Yale University, New Haven, Connecticut.

LEVENSON, EDGAR A.: The William Alanson White Institute, New York, New York.

LINCOLN, C. ERIC: Professor of Sociology and Religion, Union Theological Seminary, New York, New York.

LOWINGER, PAUL: Department of Psychiatry, Wayne State University, Detroit, Michigan.

MAZER, MILTON: Director, Martha's Vineyard Guidance Center; Assistant Psychiatrist, Massachusetts General Hospital, Boston, Massachusetts.

PINDERHUGHES, CHARLES A.: Director of Research, Veterans Administration Hospital and Department of Psychiatry, Tufts University, School of Medicine, Boston, Massachusetts.

SEIDLER, MURRAY: Associate Professor, Department of Political Science, Wayne State University, Detroit, Michigan.

SELESNICK, SHELDON: Department of Psychiatry, Mt. Sinai Hospital, Los Angeles, California.

SPIEGEL, JOHN P.: Director, Lemberg Center for the Study of Violence, Brandeis University, Waltham, Massachusetts.

WEINSTEIN, EDWIN A.: The Washington School of Psychiatry, Washington, D.C.

WEST, LOUIS JOLYON: Professor and Chairman, Department of Psychiatry, University of Oklahoma, School of Medicine, Oklahoma City, Oklahoma.

WITENBERG, EARL G.: Director, The William Alanson White Institute, New York, New York.

Part I. Historical and Biographic

PSYCHIATRY AND THE MAVERICK

by IAGO GALDSTON, M.D.

SPEAKING IN THE NAME OF Socrates, Plato in the Phaedrus states, "There are two kinds of madness, one resulting from human ailments, the other from a divine disturbance of our conventions of conduct."[1] Of the divine kind, Plato distinguishes four types, each ascribed to one of the gods. Thus the inspiration of the prophet is credited to Apollo, that of the mystic to Dionysus, that of the poet to the Muses, and the fourth, the madness of the lover, to Aphrodite and Eros.

In a previous exchange between Socrates and Phaedrus, Plato had affirmed that madness is not invariably an evil. "In reality," he stated, "the greatest blessings come by way of madness—that is heaven sent."[2] And Plato further observes, "Madness was accounted no shame nor disgrace by the men of old, who gave things their names; otherwise they would not have connected that greatest of arts, whereby the future is discerned, with this very word, 'madness' and named it accordingly."[3]

And by way of summing up that portion of the dialogue, Socrates concludes, "So let us have no fears simply on that score; let us not be disturbed by an argument that seeks to scare us into preferring the friendship of the sane to that of the passionate."[4]

The concept of divine madness was common to Greek thought. Professor E. R. Dodds, in his scholarly work *The Greeks and the Irrational,*[5] points out that it was Democritus who first talked about poetic ecstasy, maintaining that the finest poems were those composed with inspiration and by the holy breath. Democritus denied that anyone could be a great poet *sine furore.*[6] "If any man come to the gates of poetry,"

Throughout this paper the terms maverick, gifted person and genius are used interchangeably. This usage is explained at the end.

Socrates affirmed, "without the madness of the Muses, persuaded that skill alone will make him a good poet, then shall he and his works of sanity, with him be brought to naught . . ."[7]

In this connection it is of interest to recall the cryptic story of an encounter between Democritus and Hippocrates, the latter having been summoned to Abdera to treat Democritus, considered by his co-citizens to have turned mad. In the story it is related that Hippocrates came away from this encounter persuaded not only that Democritus was not mad, but that, on the contrary, he had plumbed the depths of human wisdom. Mad were those who counted Democritus mad.[8]*

Hippocrates, as is well appreciated, excluded the gods from the pathodynamics of human disease. The gods had no share in the engenderment of the so-termed Sacred Disease, nor in any other human disorder.[9] Hippocrates, in other words, was no Platonist, and in general he eschewed the philosophers and their speculations.

Hippocrates traced all thinking and feeling, both in health and in disease, to the one *organ*—the brain. "Men ought to know," wrote Hippocrates, in the oft-quoted passage from the *Sacred Disease*, "that from the brain, and from the brain only, arise our pleasures, joys, laughter and jests, as well as our sorrows, pains, griefs and tears. . . . It is the same thing which makes us mad or delirious, inspires us with dread and fear, whether by night or by day, brings sleeplessness, inopportune mistakes, aimless anxieties, absentmindedness, and acts that are contrary to habit. These things that we suffer all come from the brain when it is not healthy . . ." The corruption of the brain Hippocrates charged to phlegm and to bile.[10]

Plato's division of madness into the constitutional and the divine was not favored by the generations of Greek and Roman philosophers and physicians who came after him.

Nor did the pupil and later critic of Plato, the most learned of men—Aristotle—accept the Platonic category of divine madness. Where Plato perceived the influence of Apollonian inspiration—the order of madness that is heaven sent—Aristotle, quite as Hippocrates, saw only the effects of black bile. In the book *Problemata* Aristotle asks, and answers, the rhetorical question, "How does it happen that all those who have achieved pre-eminence in philosophy, politics, literature or art

*This story, of course, is no more than an artful conceit, one that was elaborated at a later date in the fictitious and forged correspondence alleged to have passed between Hippocrates and Democritus.

clearly are of a black bile temperament and some among them sicken because of their black bile—even as among the Heroes it befell Heracles?"[11]

In effect the concept of divine madness was not propagated beyond the time of Plato. Those among the ancients who treated of madness, of the disorder of the mind, Galen, Soranus, Aretaeus of Capadochia, Caelius Aurelinus, Celsus—considered the disorder to be caused by physical agents and conditions. There was, for those who treated of the matter, but one order of madness, that resulting from human ailments. There was no credence in any other order—and none in divine madness.

Neither Plato nor Aristotle had any followers competent to propagate and to elaborate their masters' teachings. More significantly, it was not long after the founding of Aristotle's Lyceum in 335 B.C. that Greek civilization entered upon a slow progressive decline.

In the third century B.C., Greek civilization, in the words of Professor Dodds, was entering not as it would seem on the Age of Reason, but rather on a period of slow intellectual decline which was to last, with some deceptive rallies and some brilliant rearguard actions, down to the capture of Byzantinum by the Turks. In all the sixteen centuries of existence still awaiting it, the Hellenic world produced no poet as good as Theocritus, no scientist as good as Eratosthenes, no mathematician as good as Archimedes, and the one great name in philosophy and the philosophic viewpoint he engendered, transcendental Platonism, was believed to be extinct.[12]

But as history was to demonstrate later, transcendental Platonism was not extinct but only dormant. When classical culture was reborn in the Renaissance, Plato also returned to life. Transcendental Platonism was transmuted and absorbed into Christianity. Plato's divine madness was rechristened the poets' inspiration and, for the believing, in St. Augustine's phrase, "divine illumination."

Having discussed at some length the origin and destiny of the Platonic category of divine madness, and because we are shortly to deal with its modern equivalents, it is here pertinent to establish, as clearly as we can, what Plato intended and what we understand by divine madness. One thing is certain; divine madness embraced nothing of the widely held belief that all types of mental disease are caused by supernatural interference—the belief that so tragically prevailed during the late medieval period. On the score of madness "resulting from *human ailments,*" Plato was as well informed, if not as experienced, as Hippocrates. This is impressively reflected in the book titled *Laws* (XI-934). Therein Plato wrote, "Now there are many lunatics and their lunacy takes many different forms. In the case just mentioned it springs from disease, but there is another source of lunatics who owe their madness to an unhappy native tendency to angry passion, further strengthened by ill training, a kind of men whom any trifling dissention will provoke to clamorous and scurrilous reviling of each other, conduct always and totally out of place in a well ordered society."[13]

Plato's divine madness had only this in common with madness per se, that it, too, effected a disturbance of the "conventions of conduct." But it differed from madness per se essentially in this—that it took nothing away from man. It was, on the contrary, a gift, Apollo's gift to the elect. It was the gift of Apolline mediumship "which aims at knowledge, whether of the future or of the hidden present."[14] Divine madness was thus the gift and the destiny of the gifted man; the Eternal Maverick.

Here it may profit us to conjecture why Plato was so deeply concerned with divine, with Apolline madness, why he termed it a madness and was so keen to distinguish it from the other madnesses. Most likely, this preoccupation derives from the experience of his own life, for Plato himself was, after Prometheus and before Christ, one of the greatest among the historic mavericks. "Plato's contemplative life," writes Werner Jaeger[15] "was not a gentle retirement, but was imposed by the tragic opposition of destiny to a natural ruler born out of his time." Not all mavericks take refuge in a contemplative life, but they are all, in varying measures, "out of time." They cannot be otherwise, and from this derives their divine madness.

In his Apology, Socrates pleaded that he could not but attend to the voice of his inner god, his daimonion.[16] That *was* his madness, and it cost him his life. Plato recognized the compelling force of divine madness. Plato never urged that Socrates should have behaved in any other way. He never says that the jury (who condemned Socrates) could have been wiser or better. It was, Plato affirms, inevitable for both sides to be what they were, and fate took its unalterable course.[17] This Platonic position attests to Plato's penetrating understanding of destiny's unalterable course. For given the jury, the normal, and *the maverick,* the upper variant, condemnation was inevitable. In the deepest sense of destiny, the two are interlocked. Without the normal there can be no maverick, and there is divine madness only because the others are so very sane. Yet Plato was not totally fatalistic about fate. Had he thought it completely unalterable he would not have conceived nor composed his great Republic.

There is a timeless issue in the tragedy of Socrates and it is revealed most clearly in Plato's struggle to fathom it. This issue remains unsolved to this very day, and to this very day the pursuit of its solution revolves about the Platonic ideas of *Paideia* and *Arete,* terms which are only meanly translated as education and excellence. Paideia, according to Plato, could be practiced but arete could only be cultivated, provided, of course, the potential of arete was initially given. Now, the gift of arete is the singular mark of the maverick. It is this which distinguishes him from all the rest. Socrates consistently sought out the exceptional young men, fit for the highest intellectual and moral culture, for these and their like possess quick intelligence, good memory, and eagerness to learn. Yet for all their gifts, unless they are properly educated, such men cannot attain all they want to and at the same time make others happy. This caveat should be noted clearly for herein lies the core problem of the maverick—of the upper variant; how to attain what he would in arete and at the same time make others happy; in existential terms, how to be and yet escape the hemlock cup.

Socrates had great faith in education, meaning paideia. Until he was confronted with the charges of corrupting the youth of Athens, he had no doubt but that the people, the demos, would favor the order of education he avowed and practiced. Not even when he was warned that disaster might strike him, as in the words spoken by Callicles, the Athenian man of the world, the realist in the Gorgias.[18] Socrates remained uninvolved, seemingly disbelieving the warnings. He said he did not fear "for no one who is not utterly irrational and cowardly is afraid of the mere fact of dying; it is evildoing he fears."[19]

Plato himself, be it noted, was not so sanguine either about education or the demos. When Socrates was executed, Plato, together with the other disciples of Socrates, fled to Megara.

Plato's Republic, counted with Laws, as the greatest of his works, is structured as a utopia, that is, a hope beyond the range of realization. At the end of Book IX of the Republic, when the perfect state has finally been constructed, Adimantis (who with Glaucon, both brothers of Plato, sustained the dialogue with Socrates) says, "I think it (the envisioned Republic) can be found nowhere on earth." "Perhaps," answers Socrates, "there is a pattern of it hidden up in heaven for him who wishes to contemplate it. But it makes no difference whether it exists now or will ever come into being."[20]

Plato's immediate and intimate sense of discouragement, amounting almost to despair, yet a despair fringed with some remote hopes, can be gathered from his known letters, notably the one numbered seven. This letter is addressed to the friends and companions of Dion of Syracuse. In this letter Plato comments on the execution of Socrates—"the justest man of his time,"[21] on the treacheries of the tyrant of Syracuse, Dionysus the younger, and on the assassination of Plato's intimate friend Dion. "The human race," Plato wrote, "will not see better days until either the stock of those who rightly and genuinely follow philosophy acquire political authority, or else the class who have political control be led by some dispensation of the providence to become real philosophers."[22]

Plato himself took no part in the government of Athens and, anticipating the query why he had never addressed the Assembly, he wrote—because "Plato was born late in his country's history, and the people, when he came to them, were already rather well on in years and had acquired from his predecessors the habit of doing many things at variance with his advice." He would have been delighted to advise the people he avowed, "if he had not supposed he would be risking his life in vain without accomplishing anything." "No serious man," Plato continued, "will ever think of writing about serious realities for the general public so as to make them a prey to envy and perplexity."[23]

I have dwelt this long on Plato and on his concept of divine madness because Plato was the first among the philosophers to define the singularity, the uniqueness, of the "gifted man," and to describe his plight and destiny among the demos, the commonality, the normals.

Concern with the "gifted man," with his divine madness, as we saw died out with Plato and was not revived for near to two thousand years. Why was this so? One thing comes to mind at once. To recognize the variant, one must first have a lively awareness of the normal, and for *that* one must have a sense of individuality. A sense of individuality first emerged in Homeric Greece and it attained a high degree of sharpness and refinement in the third century B.C. "In the glorious spring of the free Hellenic spirit, in those centuries full of promise that followed the age of Homer, between the dissolution of the early orders of society and the consolidation of the Attic civilization of the fifth century B.C., there came the first succession of independent personalities of European stamp . . . shadowy figures for us, but nevertheless unmistakable."[24] The flowering of Attic culture in the century and a half from the Persian wars to Aristotle and Alexander deeply affected personality formation and character.[25] There was a great redirection of intellectual interest brought about by the changes of social life in democracy, with its urgent practical problems and with the upheavals it suffered, especially after the Peloponnesian war.[26] "A new stage was reached in the conception of individuality: individual action was no longer the expression of the communal consensus, but was based on reflection—yielding the individual a firm basis and a fixed standard for his conduct in life."[27] "The individual began consciously to *use* the tradition instead of being used by it."[28]

All this cleared the stage for the variant to play his fateful role and to gain recognition as the perennial tragic character in the social drama. "In this epoch," writes Georg Misch, "choices between widely varying alternative ways of living opened out before men as never before, and the State left ample room for free activity, until, from the third century of the Christian era, the despotic rulers and the Church authoritatively stereotyped the culture that had accumulated."[29]

We had noted before that the decline of Greek culture began in the third century B.C. The reasons for that decline are numerous and mooted, but the evidence of that decline have been astutely described in the phrases "loss of nerve" and "retreat from rationalism." "For a century or more," writes Dodds, "the individual had been face to face with his own intellect and freedom, and now he turned tail and bolted from the horrid prospect."[30] Individualism was engulfed by the resurgent conglomerate mass, and it disappeared. It did not emerge again until the Renaissance. During the long interval there were men of divine illumination, but not of Apolline inspiration. Madness was of one order; the others were not mad, but possessed. Men could believe in demonology; they could not conceive of divine madness.

After Hippocrates the four most significant physicians in antiquity were Soranus of Ephesus (first century A.D.), Aretaeus (second century), Caelius Aurelianus (fifth century), and Galen (130-200). Each of these wrote on mental disorders, but none accepted the category of divine madness. "If maniacs do compose poetry or orate on astronomy and philosophy," wrote Aretaeus, "they do so not because of divine inspiration but because of their previous education."[31]

Soranus, as transmitted to us by Caelius Aurelianus, in his work titled

On Acute Diseases and On Chronic Diseases, in the section *Madness or Insanity,* first refers to Plato's two orders of madness, then cites the opinions of the Stoics and of the School of Empedocles, the latter holding "that one form of madness consists of a purification of the soul, and the other in an impairment of the reason resulting from a bodily disease or indisposition." Soranus dismisses all but that madness resulting from bodily disease or indisposition and bluntly affirms, "It is the latter form of madness that we shall now consider." He concludes his exposition with the affirmation: "Thus those who imagine that the disease (madness) is chiefly an affection of the soul and secondarily of the body are mistaken. For no philosopher has ever set forth a successful treatment for this disease."[32]

Galen, whose influence on medicine lasted from the second century A.D. up to the eighteenth century, was partisan to Plato's theories on the soul, defending them against Aristotle, but he rejected Plato's two categories of madness.[33] "The soul," wrote Galen, "is the slave of the body and is consequently dominated by the body."[34] Psychic functions, Galen affirmed, are centered in the brain and all orders of madness have their origin and causes within the body.[35]

It would not here profit us, nor is it possible under the circumstances, to survey the theoretic postulations on the causes of mental illness advanced by physicians from the time of Galen's death until the modern age. For our concern is not with psychiatric theories as such, but primarily with the concept of divine madness and of this there was no thought at all for more than 1200 years. The divine illumination of the Church fathers and the inspirations of the saints bear no relation to Apollonian madness. Nor, most important to note, does the mass of disorders that come under the rubric of demonology. None of these was divine madness. It was not until the Renaissance that the concept of divine madness revived, in part no doubt as a consequence of the rediscovery of Plato and also because of the reemergence of individuality in the Renaissance man. But now the philosophic concern with divine madness was transmuted into a concern with the source and nature of genius. Poet, painter, author, the artist in whatever medium, was deemed an inspired person—gifted if not by God, then by the latter day equivalents of the Muses.

According to Scaliger (1561), "the *dichter* creates in a godly wise." Tasso (1587) wrote: "*Non merita nome di Creatore, se non iddio ed il Poeta*—None merits the name Creator save God and the poet." Philip

Sidney, in his Apologie for Poetry (1591) compares the author with the Divine Creator.

That the issue of divine madness was thus transmuted can be readily understood for the Renaissance was singularly the Age of Genius, the heyday of the gifted, yet it soon became evident, even as it had been in ancient Greece, that there was a price to pay for giftedness, in suffering, ostracism and even death.

During the sixteenth, seventeenth and eighteenth centuries the concept of genius was elaborated by numerous thinkers and came to embrace a number of specific qualities such as ingenuity, originality, outstanding creativity. It also in time became tinged with the suspicion of madness, or at least with marked mental and emotional divergences. Vasari, in his biography of Raphael, speaks of the madness, of a something wild, disordered and fantastic, to be encountered among painters. Other writers, Ficino, Aretino, Scaliger, Erasmus, treating of other creative individuals and of genius in general, expressed similar opinions.[36]

In time two opposing schools of thought came into being, the one maintaining that the genius was an exceptional, gifted person, inspired, and a law unto himself; the other holding that genius was madness and that the genius, the so-termed gifted person was such only *because* he was mad. How earnestly each school of thought held to its opinions, how intense was their conflict, how extensive and copious were their respective literatures is now difficult to imagine. Lange-Eichbaum's work, *Genie-Irsinn und Ruhm,* proffers a monumental history and a compendium of these divergent schools of thought and the literature they produced. Lange-Eichbaum, be it noted, belongs to the genius-madness school. He denies that the genius is a *gifted* individual. The genius, he maintains, harbors no gifts, is neither genetically or biologically endowed. He affirms this in an emphatic statement: "*Genie stellt uberhaupt nichts Angeboren-Biologisches vor.*"[37] Lange-Eichbaum classifies genius as a something *Bionegative,** that is, as something sickly or sick, "*Krankliche oder Kranke.*"

The outstanding protagonist of the genius-madness theory was Cesare Lombroso (1836-1909). His initial and highly provocative work on this

*Bionegative: ein Begriff, der hier zum erstenmal gepragt wirder umfass alles, was ungunstig abnorm ist hinsichtlich der Lebensfunktionen und (oder) der Nachkommenschaft. Also Missbildungen, Entwicklungshemmungen, ungunstige Variationen (samt ihrer Vererbbarkeit) und schliesslich, als die extremste Unterabtellung, alles ausgesprochen Krankliche oder Kranke.

theme was published in 1863 under the title *Genio e Follia.* His last work on this subject was published in 1907 and was titled *Genio e Degenerazione III.* Lombroso in effect considered genius the product of a degenerative process, primarily psychological in character and mainly epileptic in nature. Lombroso's theories stirred up a great deal of controversy and psychiatrists, among others, took sides for or against them. It is of interest that Griesinger (1817-1868) and Forel, two of the outstanding psychiatrists of the period, were favorable to Lombroso's theories, while Kraepelin, Havelock Ellis and Bleuler rejected them.

The Lombroso controversy, for all its intensity, led nowhere. It did, however, have one salutory effect; it stimulated the study and publication of what came to be known as pathographies, that is, psychoclinical studies of noteworthy persons, their lives and attainments. This "literature," as we recently witnessed, can stir up deep and intense emotions, as did Freud's study with Bullit on the life of Woodrow Wilson.

It is of interest to inquire why the Lombroso controversy died out. In part no doubt because Lombroso's theories were extreme and untenable. Genius *could be mad.* Numerous pathographies reveal this clearly. But the critical intelligence would not be persuaded that the genius was such *because* he was mad, nor that genius was in itself a madness. However, it is not likely that the controversy was dissipated by a simple exercise in logic. It took something more decisive—a novel and promising approach to the problem of genius to dissipate Lombroso's arguments. This was provided by Sir Francis Galton's study of *Genius and Heredity* published six years after Lombroso's *Genius and Madness* appeared. It had been maintained that genius was not genetically transmitted, but Galton demonstrated that it was indeed in a large measure hereditary. By careful biographic studies Galton showed that there is a tendency for genius to run in families; that eminent men tend to have eminent offsprings. It should, however, be noted that Galton used the term genius in a more catholic, in a broader sense, than did Lombroso. Galton embraced in the term genius all the uncommonly gifted individuals ranging from the barely outstanding to those of the highest and rarest order.

But at the time Galton wrote his historically significant book, he had no way to measure giftedness nor any criteria by which to demonstrate objectively that the genius was indeed a genius. Galton depended on the historical datum of the individual's reputation among his contemporaries.

Galton, who was a half cousin to Darwin, was himself a good deal of a genius, and he tackled the problem of "gauge and proof" of ability with distinctive ingenuity. He sought to quantify mathematically the evidence of ability. After fourteen years of work on this problem, he published in 1883 the results of his studies in a book titled *Inquiries into Human Ability and Its Development.* This work marked the beginning of scientific individual psychology and of mental testing. "Galton," writes Boring, "believed that quantitative measure is the mark of a full-grown science and he adopted Quetelet's use of the normal law in order to convert the frequency of occurrence of genius into measures of its degree: that is to say, he set up a scale of lettered grades of genius from "A," just above average, up to "G," and to "X," which represented all grades above "G."[38] He found that "G" level was attained by only one man in 79,000 and that of "X" one in a million. Galton linked all degrees of ability in one continuum and created the means for their measurement.

The history of the development, elaboration, and refinement of psychological tests is immensely interesting, but having only a secondary bearing on our subject we may not pursue it further. This much, however, should be noted. The mental tests provided the base for a clear concept, at least in one dimension, of genius, the gifted person, the maverick.

As previously avowed, I have used the terms genius, gifted person, and maverick as if they were interchangeable, and referred to one specific order of person. In a significant sense they *are* interchangeable, and they *do* describe a unique order of person. Yet some refinement and definition is at this point desirable. Every gifted person has an element of genius about him. The capital "G" person is the highly gifted person whose giftedness, on Galton's scale, ranges between "G" and "X." The others rank lower, but irrespective of rank they are all gifted. They are likewise variants from the normal, in an upward direction. They are all in that sense mavericks. In applying the term maverick to this unique order of persons, my intention is not to signal their statistical position, but rather to focus attention on the singular psychological characteristics which carry with them distinctive and sometimes grievous psychiatric liabilities. These are the individuals, liable in Plato's terms to "divine madness," a type of madness which as my long excursus into history has shown, has been little understood and which even now is but inadequately appreciated in psychiatry.

The term maverick is of recent origin; it gained currency in 1872. It designates an unbranded creature, one that is outside of the herd. The gifted person is by virtue of his giftedness liable to exclusion from the herd, and he himself finds it natively difficult to integrate with the herd. The herd, the human commonality, finds it difficult to comprehend the maverick, feels him to be alien, and by his difference challenging and menacing. The commonality reacts to the maverick in a variety of ways. It derides him, rejects him, abuses him, and in the ultimate may destroy him, either slowly or in a burst of passion. In a great measure the behavior of the commonality toward the maverick depends upon the ways in which the maverick relates to his own giftedness in the presence of the herd. Both Socrates and Plato were mavericks, but only Socrates was put to death.

Giftedness is a heavy burden to bear and can crush the individual even where the commonality is indifferent and uninvolved. In social existence, the gifted individual needs to learn how to coexist with his own giftedness, how to reconcile his own version of *being* with that in evidence all about him. The task is exacting and hence little wonder that so many among the gifted withdraw from social coexistence to live in their private realms. It is this negative resolution of their difficulties that gave rise and appears to validate the genius-madness formula.

The gifted individual is "afflicted-blest" with divine madness, for in varying ways in different degrees, he perceives, comprehends, and relates to his environment and to the flow of experience in ways different from those of the normals in whose midst he is placed. His differences are not merely quantitative, but even more emphatically substantive. He perceives not so much more of what is the shared vision, but a something different. He is like the fabled child who saw the king not raggedly dressed but stark naked, while all the others saw him dressed in finery.

This madness of the maverick *is divine,* and it is incurable. The maverick can learn to live with his madness and indeed to profit in it for himself and for the larger group. But he cannot abdicate his madness save by self destruction.

Since the maverick cannot but be a maverick, what are the possible consequences of his interactions, as a maverick, with the overwhelming larger mass of normals? It is to be assumed, as is most commonly the case, that the herd is not favorable to him, is on the contrary hostile, and hence will make sport of him, abuse him, reject him, and otherwise injure him. The maverick's reaction, especially during his formative years, is likely to be one of three possible choices: He may withdraw from the group and become an isolate. He may "fight" the group in an endless, indeed lifelong battle (as did Schopenhauer and Nietzsche), becoming in the process increasingly hostile, cynical and misantrophic. Or he may gain a measure of acceptance among the crowd by playing the clown, a resolution masterfully executed by Bernard Shaw.

The maverick during his initial efforts to articulate with the mass may serially try each of these patterns before settling on one. I suspect that many among those who cannot but withdraw and become isolates take refuge in schizophrenia. To paraphrase this in a more telling version, I am persuaded that among the schizophrenics there are numerous mavericks whose giftedness, and the world about them, proved more than they could fuse into a life worth its burdens.

But the crucial question must be asked—is all this inevitable? Are there no other ways? Is it not possible for the maverick to have a rewarding existence among the herd, even though he does not bear its brand mark? This you will recall was the question Plato asked, and seemingly answered in the negative. Given Socrates and the jury (the demos), he said condemnation is inevitable. It is fate! But as we need recall, Plato did not blot out all hope. There still was paideia and arete. We who have the advantage over Plato in that we can scan a longer stretch of history, know that not everywhere nor at all times was the maverick uniformly rejected and persecuted. There were times of relative tolerance for the maverick, as in fifth century Athens and during the early Renaissance, when, indeed, the maverick was prized as pioneer, explorer and innovator, the avant garde of human culture. But history also shows that such periods of tolerance are rare and of short duration. Apart from this there is the germinal problem of the gifted man living effectively with his own giftedness in the world which even when it is tolerant, nay, more, is receptive, still stands out in stark contrast to the perceptive and inner life of the maverick. This contrast to which the maverick is subject and constant witness is in itself a great challenge.

Madness is the special preserve of the psychiatrist. Does then the divine madness of the maverick fall within the psychiatrist's field of concern, or is it to be left to the sociologists, the esthetes, the educators? The question is vain, for on closer examination it is clearly evident that the choice is not as open as all that. In the end, and frequently before that, many among the mavericks *volens nolens* come face to face with psychiatry and the psychiatrist. Some come as bewildered patients at odds with the world, others are embroiled in serious controversies, in futile and misdirected revolts, and still others are in conflict with the law. An increasing number, I find, are parents, themselves mavericks, who are bewildered by the giftedness of their children and do not know how to deal with them.

But then what results when maverick and psychiatrist confront each other? The results, I am afraid, are not always good and, indeed, sometimes are very bad. For the psychiatrist, trained to deal with abnormality, is chiefly concerned with the lower segment in the distribution-curve of normality, that is, with the lower variant. Classically and traditionally, the psychiatrist's efforts are directed to the bringing, or lifting of the variant, back to the normal. Furthermore, the psychiatrist shares and suffers with the rest of the normal world the tyranny of Reality. The psychiatrist is neither trained to nor is he prone to ask "What reality? which reality? whose reality?" There is a risk in asking such questions, for the inquiry may lay the questioner open to the suspicion that he is himself a bit mad. Witness the fate of such as Trigant Burrow and the younger Wilhelm Reich, who did ask such questions! But such questions *are being asked* by philosophers and by the literati. However, philosophers and men of letters do not treat patients. They can only mirror the problem, they cannot resolve it. Neither can the psychiatrist, entirely and all by himself. Still the problem falls squarely and inescapably to his lot, for as noted before, in the last analysis the psychiatrist is called on and is required to deal with the painful and wretched consequences of its nonresolution—as friend and advisor to the courts of law, as keeper to the incarcerated in the mental hospitals, and more frequently as guide and sustainer to those wrecked in their efforts to be themselves and yet to live with the others.

We need to be forthright in this matter; nothing in the present-day training or education of the psychiatrist, psychoanalyst, or other, prepares or qualifies the psychiatrist to deal with the primary, basic "abnormality" of the maverick—his giftedness—his divine madness. At best the psychiatrist can deal only with the consequent or secondary problems, and even there not much better than symptomatically. At best the psychiatrist may prove *himself* tolerant and permissive in the posture which, if verbalized, would sound thus—"Well, of course, if you choose to be that way, it's *your* affair and it's all right with me." Still there is an unspoken "but" in this posture which the patient is not likely to miss. It runs thus—"But, of course, if you had any sense you'd give up your nonsense and become a real fellow. The reality, you know, requires, etc., etc."

Mavericks understandably are not fond of psychiatrists. They shy away from them. When they are artists, they are fearful, unreasonably and erroneously so, that therapy would interfere with and perhaps even deprive them of their artistry. The maverick may be ill and unhappy, but with a healthy instinct he does not want to be made normal. His authentic psychiatric need is to be explained to himself, to comprehend himself, and in that knowledge to learn how to achieve arete in himself and yet give pleasure to the others. It can be done. But the maverick cannot be helped to understand himself in the parlance and dynamics of the oedipus complex, castration anxiety, orality, or id drives. Nor can his problem as a maverick be illuminated in terms of organ inferiority or the sublimation of unfulfilled infantile wishes. His basic difficulty derives from his being a maverick, and this conditions and generally aggravates all the problems which he encounters in the adventures of growth and development. But it is his condition of *being a maverick* and the consequences thereof that require initial illumination and insight. Without that, treatment of the rest is most difficult, if not impossible. In the paraphrase of a story told of Freud, the patient needs to be helped to understand that he doesn't *suffer* from a superiority complex—he is superior—and that this is no mean "disability." Understanding this, the maverick can also learn to understand and to tolerate the differences between himself and the rest, to understand and make allowances for the suspicions, antagonisms and jealousies he is likely to engender in the less gifted. He may even learn how to anticipate them and thus avoid or evade them. In such understanding he may gain tolerance, patience and humility, thereby increasing his arete—and also bringing pleasure to others. In learning this for and about himself, he may also apply this knowledge in the paideia of his children, for mavericks are prone to produce mavericks.

How is the psychiatrist to achieve all this with his maverick patients? First and foremost, by becoming aware of, and thus learning to recognize, this order of patient, not as yet listed in the standard classifications, and then by understanding in depth the dynamics of the maverick's intra- and interpersonal difficulties. The rest in understanding and therapeutic procedure should not be difficult to gain and master.

I want to end my exposition with an *apologia*—apologia meaning not excuse but explanation. I have given no case material either in the development of my theme or in illustration and support of the theories advanced. That was not for my lack of such materials, but because I could not include it, the space for this presentation being limited. I chose rather to present an historic survey, and that for the reason that "the proper care and treatment of the maverick" is not only a challenge to psychiatry, but is in effect of vital importance to mankind, to the maintenance and advance of civilization and culture. History shows that society has been cruelly intolerant of its upper variants. It may be kind and rewarding to its persons of talent, but not to the upper variants who challenge society's cherished values and accustomed ways. The annals of history are replete with the records of gifted men and women whom society would not gladly harbor, but instead rejected, persecuted, and destroyed. Was not Aristedes, surnamed the just, banished from Athens precisely because he was just? In exiling the law giver Hermodorous, the friend of Heraclities, the Ephesians decreed this, "Let no one be best among us, or if one must be best, let him be elsewhere and with other people." In our times we are aware of the receptions given to Darwin and to Freud. These prevailed against great odds. But who is to tell what numbers of such like did not prevail, but succumbed? Society could never afford to be so profligate with its gifted members, and, can afford it even less so now.

And who shall bring this to the awareness and before the conscience of mankind if not the psychiatrist? In the courts of man's judgement the maverick needs a pleader and defender. Let it be the psychiatrist.

REFERENCES

1. PHAEDRUS: 265. From the translation of R. Hackforth *In* Hamilton, Edith and Cairns, Huntington, (Eds.). The Collected Dialogues of Plato, Bollingen Series, LXXI, New York, Pantheon Books, 1961.
2. *Ibid.*
3. *Ibid.*
4. *Ibid.*, 245-b.
5. DODDS, E. R.: The Greeks and the Irrational. Berkeley and Los Angeles, University of California Press, 1964, p. 64-101.
6. *Ibid.*, p. 82.
7. PHAEDRUS: 245-a.

8. For an interesting rendition of this fictional meeting of Hippocrates and Democritus, *See* Boisessette Gaston: Life and Teaching of Hippocrates.
9. The author of the Hippocratic Prognostikon, however, seems to believe that certain diseases have "something divine about them."
10. Hippocrates: The Sacred Disease XVII-XVIII. Loeb Classics, translated by W. H. S. Jones, New York, Putnam's Sons, 1923, p. 175.
11. Cited in Leibrand-Wettley: Der Wahnsinn. Munchen, Karl Alter Freiburg, 1961, p. 81. Cicero in Tusculanum describes as melancholics Heracles, Lysander, Aias, Bellerophon, Empedocles, Plato, Socrates, and others, notably among writers.
12. Dodds, *op. cit.*, p. 244 (phraseology slightly modified).
13. Plato, *op. cit.*
14. Dodds, *op. cit.*, p. 69.
15. Jaeger, Werner: Paideia. Translated by Gilbert Highet, New York, Oxford University Press, 1945, Vol. 2, p. 83.
16. Plato: Apology, 40-a.
17. Jaeger, *op. cit.*, p. 73.
18. Plato: Gorgias, 486 b.c.d.
19. Plato: Gorgias, 522 d.e.
20. Plato: Republic, Book IX, 592 a.b.
21. Plato: Letter VII, 324 e.
22. Plato: Letter VII, 326 e.
23. Plato: Letter V, 322 b. and Letter VII, 344 c.
24. Misch, Georg: A History of Autobiography in Antiquity. Cambridge, Harvard University Press, 1951, Vol. 1, p. 67.
25. *Ibid.*, p. 95.
26. *Ibid.*, p. 100.
27. *Ibid.*, p. 100.
28. Dodds, *op. cit.*, p. 237.
29. Misch, *op. cit.*, p. 183.
30. Dodds, *op. cit.*, p. 246.
31. Semelaign, A.: Etudes Historiques sur l'alienation mentale dans l'antiquite. Paris, 1869—cited by Zilboorg and Henry: A History of Medical Psychology. New York, W. W. Norton and Co., 1941, p. 75.
32. Drabkin, I. E.: Caelius Aurelianus on Acute Diseases and Chronic Diseases. Chicago, University of Chicago Press, 190, p. 535-541.
33. De Dogm-Hipp. et Platon III, 7, Book V, p. 337. *In* Darenberg, Charles: La Medecina Histoire et Doctrines. 2nd Ed., Paris, Didier et Cie, 1865, p. 81.
34. Zilboorg: p. 91, A. Semeligne, *op. cit.*, p. 189-190.
35. *See* Sprengel, Kurt: on Galen, Versuch einer pragmatischen Geschichte der Arzneykunde, halle, 1823.
36. Lange-Eichbaum, Wilhelm: Genie-Irsinn und Ruhm. Munchen, Ernst Reinhardt, 1928, p. 30.
37. *Ibid.*, p. 433.
38. Boring, Edwin G.: A History of Experimental Psychology. New York, The Century Co., 1929.

Discussion by Edgar A. Levenson, M.D.

We are all indebted to Dr. Galdston for his erudite, elegant exegesis of a behavioral pattern known to the ancient Greeks but largely ignored in our more rationalistic age. We have been presented with a necessary caveat. Bound by our Kraepelinian nosology, we may fail to understand that what we dichotomize as psychopathology *in* a talented person may be more simply integral behavior; the bizarre activity being as much part of the creative experience as clucking is part of the hen's egg-laying process. Since, in this case, art simulates madness, the unperceptive family, society, and therapist isolate the creative person, mislabel his behavior as "crazy" and bid fair by their disapproval to convert it into just that—craziness.

Significantly, Dr. Galdston chose a platonic concept as his paradigm. Classical Greek is a language admirably suited to the expression of fine shadings of affect. According to the Whorf-Sapir hypothesis, if one wishes to talk about snow, talk with an Eskimo; he has the language for it. If one wishes to explore the subtler tonalities of lust, love, passion, madness and ecstasy, talk with a Greek, preferably ancient—the Greeks had a word for it! English, alas, is a more pragmatic language, suspicious of any irrational behavior, and without the lexical richness necessary to deal with it. Consequently, Plato's "madness" suffers in translation. Madness, (*μανια*, mania) meant an inspired frenzy, quite different from craziness. The Greeks had other words for insanity. It would have been quite clear to Plato, who it was suffered the divine madness and from biliousness. It would, I suspect, be clear to our contemporary hippies who have invented their own language of affect. They, too, are much concerned with the phenomena of transcending reality. "Blowing your mind," for example, is a concept parallel to divine madness; it certainly does not imply insanity although that may occur as an epiphenomenon.

This brings me to my greatest difficulty with Dr. Galdston's paper. How shall I know divine madness when I see it? How is Augustinian divine illumination distinguished from Apolline divine madness? For all of Getzels and Jackson's[1] and Terman's[2] studies in genius, we have really no typology of the upper variant. Of the phenomonology of genius, we know even less, and what we do know comes almost entirely from general literature, largely autobiographies. The use of terms like maverick, genius, and upper variant interchangeably is a consequence of this unclarity. These are terms of widely varying reference and levels of abstraction. Maverick is a metaphor of social behavior; it describes an outsider, unbranded or motherless. One can be a maverick without the slightest show of talent. Genius is a measure of intellectual capacity. There are geniuses who fulfil Dr. Galdston's criteria for divine madness, but there have been geniuses who were models of social propriety and ordinariness. Indeed, geniuses can be remarkably banal outside of their area of creativity. In a word, I believe Dr. Galdston uses this scatter of disparate terminology because there is not appropriate language to handle his concept. I am then left unclear as to whether Dr. Galdston meant to attribute some degree of divine madness to all upper variants; encompassing mavericks, geniuses, rebels, or simply original creative people. Or, does he imply that there is a special rubric, a Divine Madness limited to a relative few inspired by the gods?

At the College Treatment Unit of the William Alanson White Institute of Psychiatry, Psychoanalysis and Psychology, we have screened several hundred students over the past five years. Their behavior varies from conventional to bizarre. They exhibit talent and psychopathology in every imaginable ratio. There are certainly many students who are both talented and disturbed. But, there are rare occasions when one cannot help but understand that one is seeing, not a talented person with problems, but someone driven by the creative impulse, as out of conscious control as an epileptic. Isak Dinesen called it, "that doom of genius—the individual's powerlessness in the face of his own powers".[3] Ned Rowen, a modern composer, writes in his *Diary*, "the compulsion to compose was with me, until the age of twenty-eight, as intense as the compulsion to drink—and I was helpless before both: music flowed out of my body like a sweet but slightly sick liquid."[4] Drunk, sick, emotionally crippled, these people create relentlessly. The merely talented person comes for help often because he cannot work; the divine victim cannot live without working. The sense of possession, the almost ego-alien quality of the drive to create, the bizarre imbalance between one immensely hypertrophied function and the rest of the person characterizes this syndrome. The artist in Joyce Carey's *The Horse's Mouth* would be a prime example.[5]

Before we can even consider the case reports Dr. Galdston omitted we must agree on a definition, if not description, of this rubric. I would like to see the talented, creative maverick, or whatever, excluded. There is no doubt that they are misunderstood and stunted by their milieu, sometimes perhaps driven mad by it. But if Dr. Galdston is serious about his category, not simply using it metaphorically, then divine madness must be viewed as a gift, a possession. Plato has Socrates say that prophecy and madness share the same Greek noun. Clearly the implication is of a rarely dispensed gift. The possessor of divine madness may be then the consequence of some unknown genetic or environmental process or miscarriage, a rare human ambergris as poorly understood and perhaps as inaccessible to our ministrations as his dark brother, the idiot savant. I can only echo Dr. Galdston: we should indeed be the pleader and defender of this group. But can we treat him? Is he inexorably bound to the rock of his uniqueness? I can't think of any greater tribute to Dr. Galdston's paper than to say that he has raised many more questions than he has answered. I look forward to his further observations in this area.

REFERENCES

1. Getzels, J. W. and Jackson, P. W.: Creativity and Intelligence. New York, John Wiley and Sons, 1962.
2. Terman, L. M., et al.: Genetic Studies of Genius. Vol. I-V. Stanford, California, Stanford University Press, 1925.
3. Dinesen, Isak quoted by Rorem, N.: The New York Diary. New York, George Braziller, 1967, p. 129.
4. *Ibid.* p. 129.
5. Carey, J.: The Horse's Mouth. New York, Universal Library, 1957.

DISSENT OF NORMAN THOMAS

By MURRAY SEIDLER, Ph.D.

NO OBSERVER OF Thomas as a child or even as a young adult could have predicted the political role that he would one day play; yet to fathom America's leading Socialist, one must take into account his formative years. The mold for these years was fundamentalist Protestantism from which he has long been estranged, but apart from which he cannot be understood. For better or for worse, his was a Christian road to socialism. Thus, he "was not born or reared a Socialist."[1] In most ways, his boyhood experiences were typically those of a Protestant minister's son who was brought up in a small midwestern community during the closing years of the nineteenth century.

Born on November 20, 1884, at Marion, Ohio, he was the eldest of the six children of Emma Mattoon and Welling E. Thomas. Until his graduation from Marion High School in 1901, he continued to live in the city of his birth. The atmosphere of the home of his childhood was distinctly middle class and fundamentalist Protestant. His father and both grandfathers were Presbyterian ministers. His paternal ancestors arrived in America during the 1860's and his maternal ancestors arrived here during the 17th century.

The Thomas family lived a frugal life but they never knew desperate poverty. Perhaps the family's most important possession was an adequate library, though Norman was to remember it as being "too long on old-fashioned, theological books."[2] Nevertheless, in a home where books were treasured, he established a lifelong habit of omnivorous reading.

There is no evidence that would support the kind of theory—so often advanced in interpretations of radicalism—that Norman Thomas was drawn to social reform politics as a way of easing neurotic conflicts arising from his home situation. There was no marked revolt against the authority of his father or mother. Indeed, his parents enjoyed an unusually harmonious relationship with each other and their children. His early recollections are not only devoid of bitterness, but are those of one who "adored" his family.[3] To an unusual degree he felt love, respect and admiration for his parents.

Both parents were deeply immersed in community affairs and gave serious thought to the great world issues and problems of their time. They were also concerned about problems of a lesser magnitude. Their stern Puritan code would not permit dancing, card playing, or Sunday frivolity. But according to their oldest son, they had "a fund of kindness and tolerance for the departure of people in general from their orthodox moral and theological creed that they did not expect to need for their own children."[4]

Emma was inclined to be more liberal in her theology than her husband; yet, rather paradoxically, she was later more concerned at Norman's changing religious outlook than was her husband. The probable explanation is that she feared that she might be carried along in her son's direction, whereas Welling, who was more secure in his orthodoxy, felt no such apprehension. Her son's abandonment of orthodoxy in religion doubtless concerned her more than his embracing of socialism. She could understand abiding by a principle, but the acceptance of card playing was beyond the limit of her tolerance.

The Thomas family had no real perception of the fact that Norman would one day become so prominent. His physician brother, Evan W. Thomas, has put it this way: "We weren't supposed to become famous, we were brought up to be good. Success was not a goal in our family."[5]

After graduation from Marion High, Norman enrolled at Bucknell University but it had long been his dream to attend Princeton, then America's most important Presbyterian institution of higher learning, where both of his grandfathers and father had attended seminary. A loan from a prosperous uncle enabled him to spend his last three college years at Princeton.

He was a remarkably brilliant student placing in the "first group"—the equivalent of an all A record—in every subject that he took during his Princeton years, a record which his children always viewed with some discomfort. He graduated as the valedictorian of the class of 1905. His extracurricular involvements included debating, orchestra, and singing. Raymond B. Fosdick, later to be president of the Rockefeller Foundation, was among Thomas' close college friends. Fosdick remembers Norman Thomas as being fundamentally more conservative on practically all questions than he. "Theologically," reports the man who was to manage millions of Rockefeller dollars, "I was far more radical than he, and it never occurred to me in those days that a time might come when his position on anything would be further to the left than mine."[6]

Any account of Norman Thomas at Princeton would be incomplete without mention of his political science professor who was destined to come President of the United States. "No man made a deeper impression upon my youth than Woodrow Wilson," writes Thomas. "I was his admiring pupil at Princeton. I even tried unsuccessfully, to model my public speaking on his."[7] Nevertheless, Thomas questions whether his former professor was temperamentally fit for high political office in a democratic society. "To disagree with Wilson," reports Thomas, "was to take a great risk."[8] "All his public life he was inclined to take strong opposition or criticism as a sin against the Holy Ghost. He was sorely hurt by it."[9]

Upon graduation from Princeton, Norman followed in the tradition of his family and became involved in church work. His ultimate goal was ordination in the Presbyterian ministry. First, however, he went to New York and took a job at the Spring Street Presbyterian Church and Settlement house, which was located in a very poor and tough neighborhood. For more than a decade, he ministered to men's bodies and souls in poverty-stricken areas of New York.

First hand experience with urban poverty apparently had a tremendous impact upon this sensitive and extraordinarily intelligent young man. Much of what he saw seemed far removed from the theories he had been taught by his Princeton professors of politics and economics. Ultimately, his Christian social work experiences convinced him "that our various reform efforts were like bailing out the tub while we kept the faucet running."[10]

During these years Thomas was attending the Union Theological Seminary and became an ordained Presbyterian minister in 1911. He became very familiar with and was deeply impressed by the writings of the leading Christian socialist of that period, Walter Rauschenbuch, who taught at the Rochester Theological Seminary. Rauschenbuch argued that the teachings of Jesus Christ could not be reconciled with the materialism of capitalism or the brutality of war. At the time of his ordination, Thomas already had deep misgivings about America's domestic and foreign policies. Ultimately, he was also to become a bitter critic of many aspects of organized religion. He criticized the church for merely going along with the status quo, and argued in effect, that it should become a basic agency for the regeneration of society. In the following words, written in 1917, he sounds very much like the radical Christians of the 1960's:

In that social order, the church has been in general, a complacent partner, yet she wonders that men reject her, and, alas, the Master whose name she bears. She ignores, tolerates, or at best, only here and there openly fights conditions which warp and twist the lives of men and women and little children. . . . The church must live and teach as the constant witness to Christ, the Savior of the individual, and the regenerator of society. The two tasks cannot be separated, but must be forever united in Jesus' conception of the Kingdom. The church, then, as His witness must confess with shame, to the blindness which has made her so slow in denouncing slavery, so tolerant of war, so indifferent to child labor long after the unions were fighting it.[11]

During the years of World War I, Thomas moved in the direction of Socialist Party membership. Perhaps it is psychiatrically significant that he apparently had the need to explain his evolving political and social views to his mother. On November 2, 1917, he wrote her a long letter. Among other things he said:

I am not a member of the Socialist Party, but as I have told you I agree very nearly with very many of their fundamental doctrines. . . .

As for the church, I believe I am doing her the highest service by proving, or trying to prove that she is catholic enough to make room for social radicals. The strife between radicalism and conservatism is the battle of the future. Will it be fought by violence or more peaceful means? If the church can speak now she may do much to save the future. For myself, I believe that the Christian ethics are impossible in the present order of society, and that every Christian must desire a new social order based on cooperation rather than competition, in which the highest ideals can be involved.... Socialism without Christianity is not to my mind, the hope of the world, but you cannot Christianize radicalism by standing aloof from it.[12]

Thomas was fully prepared to risk all for the sake of his socialist and pacifist ideals. He continued:

Whether the church will tolerate me or not is for it to say. I hope it will. If it does not it will injure herself far more than it will me, and I say this without conceit, because her attitude toward me will be symbolic of her outlook for the future. I am perfectly aware that the stand I am taking is costing me many friendships, every possible chance of getting another church if I should leave here, and perhaps, though I think not, my present position. I can only say that if it costs me far more than I fear, it is to my mind the only way, for I could not maintain my self-respect and do differently nor could I serve the Kingdom of God as effectively.[13]

In October, 1918, Norman Thomas joined the Socialist Party. In his letter of application, he stated that he believed "in the necessity of establishing a cooperative commonwealth and the abolition of our present unjust, economic institutions and class distinctions based thereon." Moreover, he declared that he had "a profound fear of the undue exaltation of the state and a profound faith that the new world must depend on freedom and fellowship rather than upon any sort of coercion whatsoever."[14] At the time of writing this letter, Norman Thomas was no youngster. He was almost thirty-four years old. He did not embrace socialism out of some youthful enthusiasm, or out of some neurotic compulsion to be unconventional. He counts himself as "an exception to the good rule" that "he is no good who isn't a Socialist before he's twenty-one, or if he is after he's thirty."[15]

It is significant to note that even during these early years, Thomas did not confine his social reform enthusiasms to specifically socialist causes. In the World War I period, he emerged as a leading figure in the Fellowship of Reconciliation, a religious pacifist organization, and founded and edited its official magazine, *The World Tomorrow.* Along with Roger Baldwin, he helped to bring into being the Civil Liberties Bureau of the American Union Against Militarism. Ultimately, it became the American Civil Liberties Union, which has performed so admirably for American democracy.

During the 1920's, Thomas was an editor of *The Nation* magazine and was also a co-director of the League for Industrial Democracy from 1922 through 1937. The League for Industrial Democracy, by the way, gave rise to the Students for a Democratic Society, which has had so important an impact upon university student life in the contemporary period. In more recent years, he has been one of the most influential figures in the development of the National Committee for a Sane Nuclear Policy.

Thus Thomas has always been a broad-ranging social reformer. During more than fifty years of social crusading he has been involved in a staggering number of causes and organizations, and, in thought and action, he has rather consistently bridged the worlds of liberalism and radicalism.

What manner of man was to assume the mantle of Eugene Victor Debs? What attributes did he possess and develop for political leadership that would have to be exercised in a world with problems and challenges infinitely more complex than those that confronted his predecessor? Over the last fifty years, Thomas has become even more of an institution than Debs had been. As America's "Mr. Socialist," Thomas has remained, realistically speaking, without a competitor as the leader, interpreter and public advocate of socialism. Paradoxically, despite his unsurpassed record of losing six presidential elections, Thomas was probably as well equipped for the holding of high office as any other American of his time. He has repeatedly said that he would rather be right than be President, but would, more often than not, quickly add that he was perfectly willing to be both.

Thomas has the bearing of a leader. He is over six feet two inches tall, conveying the impression of towering size, which has been a distinct political asset. He exudes confidence and always appears to be thoroughly in command of himself, and the situation at hand. His self confidence, however, does not suggest arrogance.

He can communicate warmth and friendliness to widely varying types of people. The handshaking art of politics, so indispensable to effective political leadership, comes easily to him because he likes people and is interested in the problems of individuals as well as those of mankind en masse, although he is probably more keenly sensitive to the problems of society than to problems confronting individuals. He is interested in the masses, but is never quite of them. He has an abundance of charm and an unfailing sense of humor which endear him even to persons who meet him once or know him only slightly.

Physical energy—almost an absolute requirement for a successful political career—is something that he always possessed in phenomenal abundance. As a political campaigner, he exhibited seemingly endless vitality and stamina. The consensus among various persons who toured with him at campaign time, many of whom were twenty or thirty years his junior, is that they found it impossible to keep up with his pace. Even in his seventies and eighties, this irrepressible dissenter has been willing to talk politics with eager young socialists until the wee hours of the morning. Roger Baldwin, a nature enthusiast and long-time director of the American Civil Liberties Union, has complained that Norman even insists upon political discussions when in the woods. Thomas has virtually no hobbies or avocations, and very few nonpolitical friends. All of his tremendous energies have been devoted almost exclusively to public affairs. That these energies have been effectively and productively expended is evidenced by his prodigious output of work. His life has been perfectly in tune with the phrase from a hymn that he remembers hearing so often as a boy, "Work, for the night is coming."[16]

Only fantastic and incredibly efficient industry can explain the sixteen thought provoking books, the scores of pamphlets, and the multitude of articles and leaflets, the many thousand speaking engagements, the unending succession of committee meetings and conventions, and the correspondence of staggering proportions, including letters to and from many of the major intellectual and political figures of our time. All of these activities were usually carried out with the aid of one secretary plus much help from his devoted wife.

Far and away, Thomas' most outstanding attribute as a political leader has been his rare ability as a public speaker. Certainly he ranks as one of the great orators of the twentieth century in a class with Eugene Debs, Franklin Roosevelt, and Woodrow Wilson, but not surpassed by them.

Thomas is very stimulated by and reaches his greatest heights as a speaker in debate situations or addresses to large audiences. He excels as an orator to great crowds. So ingrained is the orator's role, that one British Labor member of Parliament has noted with some justice, that Thomas seems to be making a speech even when in private conversation. There is always a tone of urgency in what Thomas says. Eloquently, but with apparent ease, he communicates fiery indignation at some of the less lovely features of American life: the poverty, the materialism, the racism, and the gap between our democratic ideals and practices, but his manner and his words always convey to his audiences that they are listening to one who loves his country, who treasures its best traditions, but who is determined to make it an even better place in which to live. During a recent speech he exclaimed, "I don't like the sight of young people burning the flag of my country, the country I love. . . . If they want an appropriate symbol, they should be washing the flag, not burning it."[17]

Perhaps Thomas' greatest asset as a speaker is his ability to moralize in a fashion so persuasive and so impassioned that not for one second do his listeners doubt his sincerity. Earnestness, integrity, conviction—these qualities flow freely from his words. Sometimes, unfortunately, there are ministerial, holier than thou notes in what he says, but his marvelous and spontaneous sense of humor has been an effective antidote to this occasional indulgence in self-righteousness. Decades of political campaigning for an assortment of offices, ranging from alderman to the presidency provided excellent play for his wry sense of humor. In 1932 he asked his audiences not to hold Herbert Hoover responsible for the country's economic plight because, "such a little man could not have made so big a depression."[18] Thomas' reaction to Wendell Willkie's speech, accepting the Republican presidential nomination in 1940 was: "a synthesis of McGuffey's First Reader, the Genealogy of Indiana, the collected speeches of Tom Girdler, and the New Republic."[19] Thomas' charged that Willkie indicated agreement "with Mr. Roosevelt's entire program of social reform," and then concluded that "it was leading to disaster."[20]

Always handicapped and frustrated by the inability of the Socialist Party to buy enough radio time, Thomas reminded his listeners during a broadcast in the campaign of 1944, "I have minutes to discuss what Governor Dewey and President Roosevelt will have hours to evade."[21] Just before the close of the 1948 campaign, Thomas was presented with a diamond covered soap box upon being initiated into the Circus Saints and Sinners Club. "I think I know why you gave me this," Thomas quipped. "I am the only man in America who can stand on a platform, in fact I'm the only one with one to stand on."[22] He described Truman as asking his speech writers, "What's it going to be tonight?"[23] "Candidate Dewey," Thomas continued, "is clad each day in a pair of platitudes."[24] "Harry Truman," he declared during the course of the same campaign, "proves the old adage that any man can be President of the United States."[25]

To hear Norman Thomas, even if one does not share his political values, is to be captivated by his idealism, energy, and wit, a fact which is all the more true if the listener happens to be of college age, intelligent, and possessed of social consciousness to some degree. James Wechsler, now editor of the New York Post, reports that "he fell politically in love" as a consequence of interviewing Thomas for the Columbia University Spectator.[26] "He is the most impressive gentleman I've ever encountered," Wechsler wrote to a friend.[27] Free-lance writer, Milton Mayer also recalls his enormous youthful admiration for Norman Thomas: "I like Norman Thomas greatly, admire and respect and honor him, but once when I was young and idolized nobody else, I idolized him."[28]

The basis of Thomas' appeal to alert young people is well explained by journalist, Murray Kempton, who as a college student during the 1930's recollects that Thomas: "conveyed the feeling that there is something glorious about being forever engaged. He seemed always just back from the side of the sharecroppers or being egged by the friends of Frank Hague. In that guise he represented the only available peace of that varied tradition of the American radical about which Dos Passos wrote. The old libertarian dream of spending one's life in lonely combat against every form of enslavement, to the extent that it was not a Communist confusion, appeared to us to have no vessel but Norman Thomas."[29]

While Thomas was trying to remake the world, he also managed to be an effective husband and father. As their offspring and friends have testified, the marriage of Norman Thomas to Francis Violet Stewart proved to be a remarkably happy and satisfying one. A daughter-in-law has described it as "the most wonderful marriage I ever saw in my life."[30] Norman and Violet reared five children. The Thomas offspring are conventional people who lead conventionally successful lives. Early in his career as social reformer Thomas pledged himself to exert no pressure upon his children to embrace his Socialism or any of his other causes. In retrospect he wonders "if he leaned too far backwards."[31] Of his five children perhaps only one can be accurately called "a socialist. "Yet they are all interested participants in the affairs of their community—a fact which pleases their father greatly. All of Norman Thomas' children have warm and satisfying relationships with him. If they no longer entirely share his politics, they continue to love and respect him and regard him as a great man. Considering that he was an extraordinarily busy public figure and away from home a great deal of the time, his relationship with his children is uncommonly good.

During the years of their marriage and since the death of his wife in 1947, Thomas has repeatedly acknowledged his "debt" to her "for extraordinary happiness in difficult years and for strength to carry on. Whatever happens in these times of trouble I am convinced that there is no substitute for the kind of family life that she made possible for me and for our children."[32] Thomas has spoken of his marriage as "perhaps the only real success of my life."[33]

Despite the infirmities of old age, Norman Thomas, until recent months continued his career in a familiar fashion. Like an Old Testament prophet he searchingly examined American public policies and found much wanting, though he also found many things of which to be proud. Norman Thomas, now eighty-three and in the process of retirement, gives the impression of having a marvelous mind imprisoned in a very frail and increasingly useless body: his sight is dim, his hearing is impaired, his limbs are racked by arthritis, his heart is not good. Yet his ideas, his eloquence, his humor, his indignation at every instance of injustice, pour forth much as they always have during his more than half a century of social crusading.

The fact that he has not run for office since the presidential campaign of 1948, has not significantly altered his role in American public affairs. Thomas has not merely been an elderly socialist commentator on American society and world affairs. Somehow, he continues to have a way of being where the action is.

Thomas, in his eighties, has continued to have a special rapport with the young. He stirs university students as do virtually none of their professors. It is the style of university teachers to venture opinions whereas Norman Thomas exultantly speaks of his convictions. He maintains this continuing rapport with the young because he is sensitive to and concerned about the things that matter most to them. In the 1960's Norman Thomas feels most committed to peace, civil rights, and individual liberty. These are the causes of socially conscious youth in contemporary America.

Few recent developments in American foreign policy have so roused Thomas' indignation and deep concern as has the growing American military involvement in Vietnam. He played a leading role in the November 1965 March on Washington for Peace in Vietnam which was organized primarily by the National Committee for a Sane Nuclear Policy. He told the thirty-five thousand marchers the kind of things he has been telling audiences all over the country:

> We are here this afternoon because we are patriots anxious to recover for our country genuine leadership both for peace and democracy. . . . We are here because we believe the war is cruelly immoral and politically stupid.
>
> We fight a war where we count victories by the number of dead. We are shocked, with some reason, by the Viet Cong's atrocities, but retaliate by terrible atrocities through bombings and burnings, which regardless of the Administration's intention, indiscriminately destroy civilians not only in North but in South Vietnam. The torch of the Statue of Liberty must seem to most Vietnamese civilians as symptomatic not of liberty lighting the world but of the burning by which we hope to win a brutal war. . . . We are fighting for a corrupt and inefficient government, the latest in a long series and one which is

> not secure in any affection of its people. In the name of democracy we are killing the Vietnamese because it is better for them to be dead than red.
>
> All this we are doing to contain a Communism which will not be thus contained. Rather we aid its propaganda around the world. We help it to cover its failure and bridge the chasm between Russian and Chinese versions of communism.[34]

For Thomas, the Negro revolution has had deep personal meaning. He has extolled it in his writings and speeches and has worked for it in many ways, including involvement, under dangerous circumstances, in voter registration activities in the backwoods of Mississippi. From its inception, he has regarded it as one of the most encouraging trends of our time. In March 1960, in a letter widely distributed to newspapers across the country, Thomas said:

> What our white fellow citizens forget to their own hurt and their country's is one of the plainest lessons in history. Any race, group, or class having made some progress against great odds is bound to come to a time and a strength when it will not tolerate tortuous delay in winning fuller justice. American Negroes have lived through almost a century since emancipation from the horrible bonds of chattel slavery. In city after city, not all in the South, they are still discriminated against in respect to jobs, housing, even the right to equality of courtesy in lunchrooms and stores where they spend their money. Their patience has been extraordinary but as was inevitable they are arousing out of apathy or despair and doing what we praise in all our school books: struggling for their own legitimate rights.[35]

Social critic and novelist, Harvey Swados, has perceptively summarized the veteran socialist's place in the peace and civil rights movements:

> In each he has played the role, not just of elderly well-wisher, but of tribune and mentor. No historian will be able to interpret the civil rights battle without assessing the significance of the inspiration Norman Thomas has given to such striking figures as the organizers of the March on Washington, many of whom not only belonged to his party but continued to look at him for intellectual, moral, and practical aid, both in public sessions and in behind-the-scenes activities.[36]

Thomas finds it hard to be sanguine about economic and social trends in contemporary America. Long an advocate of the welfare state, Thomas rather ironically contemplates its emergence in American society with some misgivings. He concedes that capitalism managed along Keynesian lines "can give us an increasingly affluent society, provided it stops the mouths of protest by grants of bread and a little cake. A prosperous business community has learned to accept that fact cheerfully—especially if it goes along with tax reduction."[37] He is not at all sure that America's expanding social welfare programs—which he vigorously favors—are leading the nation toward a democratic socialist conception of a Great Society. "The welfare state," he notes, "does relieve much misery, opens many doors to a fuller life. It may be a stage on the way to a much fairer economic system with a degree of democracy in industry now lacking; this is the democratic hope. Or the welfare state may become a glorified version of the Roman Government's bread and circuses."[38]

Yet Thomas' deepest concern is not so much with the specifics of Johnson's domestic policies about which he has many reservations. He is convinced that Johnson's aspirations for American society are incompatible with his conduct of foreign affairs:

> The great trouble is with the President's foreign policy. No great society can be founded on an attempt by one mighty power to police the world, by force of arms if necessary. Certainly the revolution of rising expectations cannot be met by emulating some of Communism's bad characteristics in the name of anticommunism. Insofar as President Johnson has won a consensus on his intervention in the Dominican Republic or on his escalation of the war in Vietnam, it is a consensus based on self-righteousness and faith in violence, incompatible with any decent society—in terms even of enlightened self-interest.[39]

During the last decade the well established reputation of Norman Thomas has soared. He has become, often to his great discomfort, a kind of living monument, but a very useful and active one. He frets about the fact that he is widely and inaccurately regarded as a kind of benign rebel. Yet, realistically, he concludes: "However, uneasy as I am about being spoken of too well by too many men, I don't feel I'm obliged to go out and smash somebody's window to prove that I'm not respectable."[40]

Despite his discomfort, accolades continue to be heaped upon him. In May of 1964, the Italian government awarded him its highest decoration, The Star of Solidarity First Class, in appreciation of his aid to Italian anti-Fascist forces during earlier years. *Esquire* has included him in its list of "unknockables," persons of untarnished reputation who represented "the sum of human virtue in 1966."[41] *Life,* in a long and glowing account of his activities, has dubbed him "the Dean of Protest."[42] Even the much honored Robert Frost regarded a meeting with

Norman Thomas as "one of the great moments of my life" and indicated that he yielded to no one in the depth of his admiration for Norman Thomas.[43]

Now in the autumn of his life, Norman Thomas continues to stand out as one of the towering figures of his time. It can be argued that there are two types of political leaders. The one is the person of average talents who makes outstanding use of the capacities that he possesses. The other is the unique individual who brings to his leadership role striking gifts and brilliance. In his own career Norman Thomas has displayed qualities and powers of personality so original that he has undoubtedly earned a lasting place in the latter category.

Charisma, a favorite term in the lexicon of contemporary social scientists, seems to be especially applicable to America's most important socialist leader. Mr. Thomas is a charismatic leader in the sense that he possesses exceptional individuality that sets him apart from other men. He has unusual appearance, bearing, and voice. In contrast to so many others of undistinguished demeanor who direct the affairs of men, Thomas looks like a leader. An autonomous person with a great confidence in his own judgments and values, he is determined, almost compelled, to convince others of the soundness of his positions. He has the power essential to charisma to impress and to excite attention. Curiously, despite having been resoundingly defeated in six presidential campaigns, there is still the presidential aura about him. To be with Norman Thomas is to have the feeling that one is in the presence of greatness.

Earnest admirers of Norman Thomas often bemoan the fact he was never elected to high office. Yet, given the realities of American politics, if he had been elected, would he have been able to live by his "voice of conscience"?[44] Would he have been in a position to champion a host of unpopular causes from which even liberal office holders usually have shied away? The answer to these and similar questions has to be an emphatic "no". Norman Thomas could not have played a comparable role in American life had he operated, as doubtless would have been necessary to gain office in the context of major party politics. He would not have been permitted his allegiance to socialism, which has been the guiding thread of his career and which has set him apart from most American political leaders, whose interest and commitment is to the politics of expediency and consensus. In short, this vehement, yet strangely respectable rebel with a clarion voice calling for peace, plenty, and freedom has been able to make his vital contributions to American life precisely because he was never elected to high office.

Thomas may be accurately thought of as a successful failure. He never won an election and failed to build a strong Socialist Party in the United States; yet, to an amazing extent the ideas that he has championed for half a century have become part of the American social fabric.

Perhaps it is of psychiatric interest to inquire why Thomas, in view of his physical handicaps and age, has continued to be so completely involved? He candidly says that he wants "to go out with a bang, not a whimper."[45] He also explains that "all my adult life I have found this an interesting, if cockeyed world, in which there are causes outside oneself which richly repay devotion."[46] Norman Thomas continues to be sustained, perhaps more accurately, driven on by a compulsive humanism and idealism which is characteristically revealed in his recent speculation about man and his destiny:

> Once socialists of various shades found their utopias a satisfactory substitute for any kind of Kingdom on earth. Privately if not publicly most democratic socialists—and, I suspect, many communists—now acknowledge that they must settle for social changes which, while of immense value, cannot of themselves banish crime or answer all the questions of the human heart caught in this strange, wonderful, beautiful, terrible universe.
>
> But this lack of a utopia or any sure far-off divine event to which the whole creation moves is no reason for despair or for a let-us-eat-drink-and-be-merry philosophy. Some tomorrow each of us individuals will die. But life will go on, and life can be made better for the whole human family. We are not damned by our gods or our genes. We can find meaning for ourselves and an end to frustration and alienation by joining in a struggle which must be worldwide, a struggle for a fraternity of mankind. . . . And in man's confused record and present conduct there are elements to give us hope that such a utopia can be attained.[47]

REFERENCES

1. Thomas, Norman: We Have a Future. Princeton, New Jersey, Princeton Press, 1941, p. 10.
2. Thomas Reminiscences. Part I, Columbia University Oral History Project, p. 6.
3. *Ibid.*
4. Thomas, Norman: Autobiography. (unpublished manuscript), p. 21.
5. Interview with Evan W. Thomas, September, 1960.
6. Fosdick, Raymond B.: Chronicle of a Generation. New York, Harper & Brothers, 1958, p. 52.

7. Socialist Call. October, 1956, p. 12.
8. Thomas Reminiscences. *op. cit.*, p. 17.
9. Autobiography. *op. cit.*, p. 31.
10. Thomas, We Have a Future. *op. cit.*, p. 11.
11. Thomas, Norman: "The Church and Industry". (unpublished essay), 1917, p. 9.
12. Norman Thomas to Emma Thomas, November 2, 1917.
13. *Ibid.*
14. Norman Thomas to Alexander Trachctenberg, October 18, 1918.
15. Thomas Reminiscences. *op. cit.*, p. 43.
16. Interview with Norman Thomas, September, 1960.
17. New York Times. October 31, 1967, p. 1.
18. Thomas Reminiscences. *op. cit.*, p. 36.
19. Barnes, Joseph: Willkie. New York, Simon and Schuster, Inc., 1952, p. 195.
20. *Ibid.*
21. Socialist Call. July 14, 1944, p. 7.
22. New York Times. October 27, 1948, p. 18.
23. *Ibid.*
24. *Ibid.*
25. *Ibid.*
26. Wechsler, James: The Age of Suspicion. New York, Random House, 1953, p. 41.
27. *Ibid.*
28. Milton Mayer to the Author, August 2, 1960.
29. Kempton, Murray: Part of Our Time: Some Ruins and Monuments of the Thirties. New York, Simon and Schuster, Inc., 1955, p. 324.
30. Thomas Family Interview, September, 1960.
31. "Autobiography," *op. cit.*, p. 21.
32. Thomas, Norman: A Socialist's Faith. New York, W. W. Norton & Co., Inc., 1951, p. x.
33. Luncheon in Honor of Norman Thomas. Testimonial Booklet, February 4, 1950, New York, p. 3.
34. Thomas, Norman: A Message to Johnson. Sane World, V, January, 1966, p. 2.
35. *See* The Detroit News. March 12, 1960, p. 4.
36. Swados, Harvey: New Reasons for an Old Cause. Saturday Review. XLVII, February 1, 1964, p. 43.
37. Thomas, Norman: President Johnson's Great Society. Christian Century LXXXIII, March 9, 1966, p. 301.
38. *Ibid.*
39. *Ibid.*
40. Playboy Interview: Norman Thomas. Playboy, XIII, November, 1966, p. 96.
41. The Unknockables. Esquire, LXV, June, 1966, p. 84.
42. Wright, Sylvia: The Dean of Protest, Life, LX, January 14, 1966, p. 62.

43. REEVE, F. D.: Robert Frost Confronts Khrushchev. The Atlantic, CCXII September, 1963, p. 34.
44. *See* THOMAS, NORMAN: Great Dissenters. New York, W. W. Norton & Co., Inc., p. 166.
45. WRIGHT, SYLVIA: *op. cit.*, p. 60.
46. Norman Thomas to Bruce L. Sandberg, December 22, 1962.
47. THOMAS, NORMAN: President Johnson's Great Society. *op. cit.*, p. 303.

Discussion by Sheldon T. Selesnick, M.D.

I have fond memories connected with Norman Thomas and the author as well. Dr. Murray Seidler is a dear friend who was a former roommate in college. In those days our bull sessions which eventuated into activism on social issues were a threat to our graduation. A hopeful sign of the times is that now, almost a quarter of a century later, we have a national forum in which not only to discuss dissent, but to clarify the psychological issues on the subject. But, the "respectable rebel," Norman Thomas, even after his stroke—and we wish him well—may still resent being referred to as a statesman. When I heard him being introduced to the student body at the University of Wisconsin with that title twenty years ago, Thomas retorted sarcastically, "Look, the word statesman is reserved for politicians who have died, and I still feel very much alive."

Murray Seidler's address has this in common with his book:[1] the clarity of the exposition portrays vividly the gallantry of the courage of a vibrant man Norman Thomas. But what innervates a man of such energy and keeps him vital? What makes him a "successful failure"?

A crucial issue is for us to try to understand why Norman Thomas selected socialism out of the myriads of possible positions that he could take. Why did he not join the Communist Party? After all, socialist theoreticians could easily come to believe in the ultimate democratic-communist position. Whereas Murray Seidler may be quite correct in his major thesis that Norman Thomas' dissent was not born out of neurotic needs, he sometimes discounts Thomas's psychological motivations. In his book, for example, Seidler notes that Thomas frequently wrote to his mother in an attempt to help clarify his own views. Seidler here acknowledges that "perhaps it is psychiatrically significant that [Thomas] had to explain . . . his views to his mother." However, this significance is not elaborated, since Seidler equates neurotic motivations with psychological ones. Not all psychological motives are neurotic, anymore than all dissent is based on neurotic displacements.

In the book, *Opinions and Personality*,[2] social scientists who studied the lives of ten important personalities at the Harvard Psychological Clinic, emphasized repeatedly that the function of an opinion is a compromise between reality demands, social requirements, and inner psychological sets. Why then did Seidler's biography note the wealth of data on the reality and social demands on Thomas, and at the same time neglect material on his psychological pressures? An understanding of this enigma can be gained from the vantage point of the biographer's transference to his hero, who himself was highly skeptical of Freudian theories.

Harry Fleischman, in another biography, quotes Thomas as saying: "What a set-up for the modern psychologically minded biographer or novelist! A study in revolt born of reaction from Presbyterian orthodoxy, and the Victorian brand of Puritanism in a Middletown setting! The only trouble is that isn't what happened. I both loved and respected my parents."[3] Elsewhere, in talking about his adherence to some of his father's creeds, Thomas went on to say, "I challenge any Freudian to show that it had led to as much nervous disorder, instability or more unhappiness than the current laxity of standards.'"[4] Superficially, this may seem to be the height of naiveté on the part of a member of the sophisticated intelligentsia who denies insecurities by stating that he loved his parents. Fleischman reports that Thomas often was an "over talker"; however, although Thomas took great pride in his social successes because he had been a sickly awkward child, this is understandable and should not be tabled as neurotic.

But why then did Thomas and subsequently his biographer, refuse to examine psychological motivations? These motivations, I submit, are linked to Thomas' leading the Socialist Party instead of the Communist Party, and are uniquely connected to his relationship with his parents. Norman Thomas did indeed love his mother, who was the dominant parent in his family and, according to Murray Seidler's report, gave him only one admonition—"Never hurt your father." Yet, this son and grandson of fundamentalist ministers could not abide by the doctrinaire, authoritarian, centralistic, and dogmatic philosophies either of the Communist Party or of the Freudian movement upon which he could eloquently and freely, without discomfort, discharge his enmity. These doctrines were scorned by one who could not openly attack the orthodox standards of his father if he wished to remain in his mother's favor. Nor was it a direct affront to his father for Thomas to attack his onetime ideal Woodrow Wilson. Thomas could envision Woodrow Wilson as a Calvinistic, austere gentleman, with whom he could disagree on basic issues. It was the Calvinistic doctrine that preaches that "somewhere, somehow, people are enjoying themselves, and that ain't good." Perhaps Thomas' sense of humor, like much of humor, is not only hostility in disguise, but allows for acceptable pleasure for the speaker. Often following a Thomas witticism, one could hear the raucous zealous laughter by "Mr. Socialist" himself. The zeal itself, especially the evangelical missionary enthusiasm, must have been a source of pride to Norman's mother, whose parents were missionaries in Siam where they achieved acclaim, and were even mentioned in the musical "The King and I."

It is no wonder then, that the Socialist Party, so much in need of a leader, found the right man at the right time. Thomas could be comfortable leading a party which was largely devoid of dogmatic centralism. This leader was not a "Man of La Mancha", because he had a dream that was a possible dream. The Socialist Party platform of 1932 with its many provisions, from child labor laws to social security, thought then as anti-American, are now part of our legal heritage. In *Group Psychology and the Analysis of the Ego,* Freud pointed out that a leader has the capacity to evoke an extreme positive transference from his followers, whereas the leader has few or no attachments to anybody but himself. It is, Freud believed, this narcissistic quality that makes him a leader. Freud considered that the leader loves other people only insofar as they can serve his own needs. The leader is "of a masterly nature, self-confident, and independent."

He thus represents qualities that the group members themselves cannot attain, and therefore, he becomes their ego ideal.[6] This Freudian view of a leader, as one devoid of meaningful object-love, may be true of narcissistic individuals who are disturbed in their interpersonal relationships, but is contradicted in the life of Norman Thomas, who formed strong personal ties. When a dear friend suffered a loss, Thomas himself became deeply saddened. I am familiar with this capacity since I knew intimately a group of ten young men whose ego-ideal was Norman Thomas and who were deeply influenced by his humanistic philosophy. The father of one of these boys, Emanuel Seidler, was a close friend and socialist colleague of Thomas and another ego-ideal for this same group. Murray Seidler and another sociologist published their inner workings under the title *A Jewish Peer Group.* They pointed out that "the greater the peer group's integration and cohesion, the more will it usurp the role of the family in the socialization process. In courtship and marriage, social and political beliefs and behavior, and occupational selection, the peer group studied markedly influenced the behavior of its members."[7] One of the constantly reinforcing factors in keeping this group viable was the encouragement and warmth emanating from Norman Thomas. Professor Louis Coser (see his chapter on "A Sociologist Views Dissent") makes the cogent observation that a constructive dissenter is an individual who can remain involved yet disengaged. Norman Thomas is an excellent example of such a dissenter; his patriotism was never questioned, but his wish to make his country better led Norman Thomas to say, when he heard of young boys burning the flag, "Don't burn the flag; wash it."*

As we sat on the grass at a farm outside Ann Arbor, Michigan, some 20 years ago, to hear Thomas address a group of young socialists, an airplane flew overhead. How in the midst of plenty, Norman Thomas asked, is there such paucity, primitivity, and misery while we produce for war? Despite his plea, research departments, including our own, are curtailed because federal funds are being expended in this spiritless and senseless Vietnamese war while private enterprise spends millions for marketing research. There are established laboratories in this country which measure audience reactions to various commercials and new television programs by means of physiological and computer equipment more complex than most universities have at their disposal. With this same equipment it would be possible to validate or invalidate major psychosomatic theories that men have spent their scientific lives in trying to evolve. As Tevye in "Fiddler on the Roof" lamented, "would it spoil some vast eternal plan, if I were a wealthy man? . . ." Yet Norman Thomas could always find some optimistic note in the concept of

*Incidentally, this characteristic of the involved yet disengaged influences is precisely the sine qua non of the effective psychotherapist. The patient who feels respected can accept constructive criticisms, presented in the form of interpretations, from a therapist who allows himself to be involved and yet can be somewhat objective. The classical analyst, with his noninvolvement, makes interventions which can be felt as attacks similar to those recent social dissenters who are uncommitted to their country. Or equally ineffective are the over-involved therapists whose sympathy for their patients precludes constructive perspective. In politics these are the super-patriots.

responsible dissent. To paraphrase a passage in his *Voices of Dissent:* "Our world sorely needs a wisdom more appropriate to it than any which we have inherited . . . The merit of dissent is great, when dissent is construction and not a dissent of negative despair or brittle cynicism."[8]

REFERENCES

1. Seidler, M.: Norman Thomas Respectable Rebel. Binghamton, New York, Syracuse University Press, 1967.
2. Smith, M., Bruner, J., and White, R.: Opinions and Personality. New York, John Wiley and Sons, Inc., 1956, p. 275.
3. Fleischman, H.: Norman Thomas, New York, W. W. Norton & Co., Inc., 1964, p. 31.
4. *Ibid.*
5. Seidler, M.: Norman Thomas Respectable Rebel. *op. cit.,* p. 10.
6. Alexander, F., and Selesnick, S.: History of Psychiatry. New York, Harper & Row, 1966, p. 205.
7. Seidler, M., Ravitz, G.: A Jewish Peer Group. Am. J. Sociology, Vol. LXI, #1, July, 1955, p. 11.
8. Voices of Dissent—A Collection of Articles from Dissent Magazine, preface by Norman Thomas. New York, Grove Press Inc., 1958, p. 9.

Part II. Individual Dynamics

REBEL AND REACTIONARY, SIBLINGS UNDER THE SKIN?

By SHELDON B. COHEN, M.D.

Introduction

IMAGINE, if you will, Moshe Dayan coming into your office and for forty-five minutes pouring out his thoughts and feelings. Imagine further that in a day or two Gamal Nasser comes into your office and also speaks his mind. Think of this process as continuing for weeks and months, and you will begin to share with me an uncanny experience that was mine several years ago when I was treating two young men. Since their dissent and divergency concerned civil rights and segregation, one could have picked no better pair of polar personalities, e.g. George Wallace and Stokely Carmichael. Bringing the focus down to earth, neither patient was an international or national personality, and it is doubtful that their names are noted or remembered by more than a handful of friends. They were lieutenants in the fray, making no policy decisions, planning no overall strategy. But on the ground that was theirs they showed energy, intelligence, initiative, dedication and commitment. They participated in minor skirmishes, were present on historic occasions and like most young soldiers turned to other pursuits when the opportunity presented itself. It should be noted that both were WASPS from the right side of the tracks.

Atlanta, Georgia, in the early and middle 1960's provides an excellent vantage point to view the drama of desegregation. Robert Coles'[1] masterful delineation of children who first crossed the barrier was based to a considerable extent on work with students in the Atlanta Public School System. Atlanta is a metropolis set down in the red clay foothills of the old South with a populace that has come mostly from other places. Large numbers have come from the South's rural areas and small towns. More northerners have rooted here than Sherman might have dreamed of commanding in his March to the Sea a little over a century ago. The watchwords have been "growth and forward movement." It is the home of a Negro minister who, because of his development of the technic of nonviolence, became a Nobel laureate. It is also the home of an unknown restaurateur who with axhandle defiance of authority found the limelight and eventually the Governor's mansion. In those days the lines seemed to be more clearly drawn; Vietnam was not such a devisive issue and SNCC was composed of nonviolent students. Tension ran high at times, but by and large groups such as the SCLC and the KKK picketed and counterpicketed at will.

Don

On an October day several years ago a young man attending a local university telephoned for an appointment. Don, a rather large, hulking lad of twenty-one, ambled into the office, settled into a chair and began to talk. Of fair complexion, he was casually attired with hair and clothing somewhat dishevelled. This appearance and "uniform" were typical of the two and a half years of therapy that followed.

He had attended an Ivy League school for two years, dropped out for a year, and was now completing his junior year at a school some two hours drive from the small mountain city where his parents resided. He was seeking help because of marked feelings of general anxiety, especially in relationship to his girl friend. Almost immediately he began to describe her in a slightly detached, intellectual manner, for him a characteristic mode of relating. He considered himself to be in love with her, but expressed concern over her feelings of loneliness and anxiety. He was especially concerned because her I.Q. was 20 to 30 points lower than his.

At the time of entry into therapy, Don lived in an apartment near the campus, returning home on weekends and for brief periods during vacations. His younger sister, a student at the same institution, had similar living arrangements, and his older, married sister lived in New York City.

He was born in a northcentral city and his family had moved several times because of his father's business. The first two moves were to other cities, after which the family spent four years in rural New Jersey. There they lived in a home the father built himself. When Don was eleven the family moved to Atlanta, remaining there until after Don entered college. Therapy produced considerable curiosity about his early life and on visits home he queried his parents and would bring information back to subsequent sessions. He was told that at age eighteen months he was placed on the toilet for hours; he would not defecate but soiled his pants soon after being taken from the bathroom. At twenty-seven months he had polio and "nearly died," allegedly being saved by an out-of-town specialist brought in by the family. After that he was fearful of men who carried briefcases and of women dressed in white. At thirty-three months he was hospitalized for a tonsillectomy and left the hospital, running a mile before he was caught. He was described as a "very determined little baby." He was unable to talk until he was three and then would speak only to his sister and friends, but not to his parents or anyone else until he was five. At seven and a half, while exploring the woods, he thought his family had forgotten him and walked for five miles before being picked up by the police. He recollects thinking of death and heaven at this age with feelings of being lonely and deserted. At eight his mother took him to a psychiatrist primarily because of problems in school, where he was prey for bullies. He kept a frazzled teddy bear until he was ten.

His memories of early adolescence were primarily those having to do with sexual enlightenment and investigation. His mother was the one who told him about sex. Masturbation was accompanied by moderate guilt feelings. When he had a rash on his penis his mother examined it. When he developed hair on the chest at fifteen he didn't want his sisters or others to know about it.

Generally he did well in school and even when his grades were not the best his parents were told that he was the brightest in his class. He remembers running home to his mother with his grammar school grades. He skipped the fourth grade. He did well in a private high school.

His parents were married in their early twenties and had three children in rapid succession, Donald being the only boy. His sisters were 18 months older and 18 months younger. His mother's family had fled Russia when she was a toddler. Supposedly because of her propensities for theatrical activities she was banished from the Mennonite Church. His father was a chemist, a hardworking man given to violent outbursts of temper. Donald saw his father as a distant man who bullied him, especially during the latent and early adolescent years. Don was frightened and resentful of his father. After some of his rage towards his father subsided with therapy he saw the observations as valid, but could then also view his father as someone who would always help out whenever he was in trouble.

Don's Civil Rights Activities: In late spring, seven months after entering therapy, Don casually said, "By the way, my sister is now in jail. She was arrested two hours ago in a sit-in." For two sessions he discussed civil rights, then ignored the subject until his sister went on trial three months later. Again there was no significant mention of the topic for several months until a week before the assassination of President Kennedy. Thereafter, this theme appeared with regularity until termination of therapy.

Don thought his parents approved in principle of his sister's and his participation in civil rights activities, although his father was fearful of the impact of his daughter's arrest on his business. Although integration was accepted by most of the university, many of the students and professors in the department of Don's major interest were segregationists. Some of his shyness and awkwardness dissipated while in the circle of picketers and protesters. There was an expanded social life with people who were on his side.

On the morning of November 21 Don announced that he and his sister were going to join a group picketing downtown. He felt that this would "blow the lid off of complacent people." He expected a telephone call from his father but said that he was of age and knew what he was doing. He had purchased new clothes and was going to wear a coat and tie for the occasion. Three days later he announced it would take more than an assassination to stop the movement. He was upset by the President's death, especially when he observed others who didn't seem to be concerned. On the night following the assassination there was an involved dream in which he was trying to secure a date with a beautiful young girl dressed in blue. Despite the presence of her parents in twin beds, he was finally allowed in and "I believe I got to arrange a date." A week later there was a nightmare in which someone said, "Are you sure that you are white?" A hand was pressing against his genitals, about to snap off his penis. The associative theme was getting back at father by involvement with civil rights and father in turn retaliating with castration threats.

Don openly expressed ambivalence about the civil rights group at school saying, "I want to be part of it, yet I want to stay away." He wanted to hear both sides of questions and was among students who urged that die-hard segregationists as well as civil rights advocates be invited to speak before the group. Without any apparent planning or much effort on his part, he was elected chairman of the campus integration group. Some time later he was somewhat surprised to find that "I'm prejudiced. I was shocked at an integrated party when a good looking white girl was dancing with a Negro." At times he recognized a compulsion to enter into arguments with his conservative classmates and felt himself afflicted with "demonstration fever," a need to go out and carry signs. The tempo of his activities heightened in the last weeks of the 1964 presidential campaign. He even joined a traditional campus political group, the Young Democrats, "for the duration," donned coat and tie and rang doorbells in behalf of local candidates. A group of conservative students in the department announced formation of a Minuteman group. One sent out notice that he was going to shoot Don. Fisticuffs erupted, and for some weeks Don was concerned that this might be more than just an empty threat. After the election, civil rights activities were quiescent for Don. He married over the winter holidays and became engrossed in field work for his master's thesis, lamenting his previous neglect of school for politics.

Selma: A week before this sleepy little town was to make headlines, Don said he had been thinking of going there, but he was concerned about possible danger and the loss of time from school work. At that time he presented a dream in which he was walking down the street towards a young woman in a white dress. Since there were pickpockets around he put his wallet in a side pocket with his hand around it. A couple of Negroes came in and tried to snatch the woman's purse. He tried to intervene but slipped and fell and was lying on the ground yelling for help. The police moved in but no one else would do anything. Finally it was 12 noon on Sunday; all the people coming from church saw and stopped the purse snatchers. Don questioned his own motivations in going, "How much am I doing this for freedom? How much for the ride?" He was fearful that his departmental head would learn of his activities and disapprove, and his grades would suffer.

He went to Selma and returned tired but without incident. This was the first night he had spent away from his wife. He resented her not accompanying him but was aware that her job as a school teacher could have been jeopardized if she had participated. He was also angry at her when he came home in the middle of the night and she did not wake up to greet him. He was surprised and a little ashamed that he had feelings towards women who were present in the march, thinking that such thoughts should have disappeared since he was married. He saw the whole march symbolically as "masturbating in front of this big state capitol."

Providing a somewhat different view of Don's activities was a brief attempted therapeutic encounter with his younger sister, Joan. After she made a panicky flight to New York, Don and her father urgently requested consultation. Joan was tall, gawky, large-boned, and dressed in a plain suit. She initially displayed considerable anxiety and depression. Uncertainty, doubt, anxiety plagued her life. She was due to graduate from college within a few months and had no idea what she would do. She did not want to work in the field of her studies and thought of taking a job in a variety store. She had conflicting, disturbing feelings about two men—one a fellow student who had talked of shooting himself and another, an instructor, who also appeared to be disturbed. There were many conflicting feelings about her civil rights activities, her trial, demonstrations, and further involvement. Psychotherapy was seen as a fearful, unknown, unnatural situation, and she consented to see me only because her brother seemed so different and happy after a year and a half of treatment.

However, she was fearful therapy would make her a "well adjusted person," who would be content with a humdrum existence. Despite all attempts at support and reassurance, therapy remained an unpleasant, unreal experience. She missed a number of sessions. Therapy was terminated at her insistence after eleven sessions.

Joan had many vivid memories of her parents' fights and her father's fierce temper. She noted that Don had similar traits which had diminished considerably with therapy. Joan had very angry feelings towards her mother who seemed to consider herself infallible. Joan had always felt left out. "I was number three; Barbara was so good and Donald was Donald and Joan was irresponsible but we love her anyway." During adolescence there was a flurry of daredevil activities; e.g. with a girl friend she would sneak out at night to ride horses. She took pride in telling me that Don rode the horse but once and her older sister was petrified of it.

During one of her last sessions she confessed having taken a paperback book from the school book store a couple of days after a lecture on juvenile delinquency. This was the first stealing in a long time and on this particular occasion she had plenty of money and bought other items. Although there was a feeling of shock afterwards, "part of me was happy about it." She had a fantasy that it would be simple to give in and take one book each day. By the end of the quarter she would have an extensive library which she wouldn't have time to read.

During the first two years of college she was interested in her sorority and had a good friend who was president. After a while sorority activities became silly and meaningless, the other girls all seeming so much younger. Since it would make a big stink around campus if she dropped out, she continued to belong but was a nonparticipant. During the two years of her sorority activity she and her cohorts were known as the intellectual rebels on campus. As the sorority became more like other groups her interest waned. At the same time she began going to civil rights meetings, picketing, demonstrating, ultimately getting arrested. She felt her activities had not alienated anyone at school since it was not fashionable on the campus to be a segregationist. Civil rights activities were seen as drawing people together. Those involved generally did not belong to fraternities or sororities. She respected the segregationists who would go out and fight for their convictions much more than those who espoused no opinion. She was annoyed with the mother of a northern governor who flew down, had herself jailed, paid a fine, and went home, saying "She ought to have stuck around."

Carey

Several years previously when Carey had displayed considerable anxiety and difficulty in forming relationships with peers and when his grades began to fall, his parents had been advised to seek psychiatric care for him. They demurred on grounds of potential damage to his aspirations for a political career. However, when in his 16th year he was discovered to have been the recipient of pornographic material over a long period of time, recommendations for psychotherapy were rapidly followed up. At that time he was a junior in a private day school. His parents expressed concern over his sexual identity, noting that he had shown no interest in girls, that he didn't care for sports or other masculine interests and had always been an unusual child. One of his mother's first remarks was that he always wanted to be completely different from everyone else. At three he removed knobs and locks from all the doors in the house with a screwdriver and his parents considered this a sign of intelligence. At six, when he was constantly stopping up toilets, his behavior was no longer thought to be laudable—particularly when punishment was ineffectual in producing change.

Carey was a short, scrawny, bespectacled lad whose predominate mood was one of anxiety, tinged with feelings of anger at being "put upon." Details of his life poured out rapidly. Almost immediately he characterized himself by saying. "I'll argue with anybody about anything." He readily accepted the the idea of help, as had his mother. Throughout therapy his father was quite leery of the whole process, especially when closer involvement of parents was suggested. Early in treatment there was a hiatus of three months caused initially by his father's insistence that he stop, and later prolonged by a bout of nephritis. An hysterical reaction while in the hospital convinced the father that "some good might come out of treatment."

Along with his parents and two younger sisters Carey had spent his entire life in a quiet, exclusive, suburban neighborhood. He was resentful of Barbara, two years his junior, who was popular and athletic. He was fonder of Cindy, six years younger, but picked on her.

His father ran a prosperous manufacturing establishment to which he had fallen heir. In recent years the father had complained of increasing difficulty, ostensibly because of labor union demands. A shy, tense man of short stature, he gave the impression of just holding on, that the struggles of family and business were almost too much for him. He suffered from a "spastic colon." He assiduously avoided conflict and violence but frequently flared up with short-lived, generally unprovoked rages at wife, children, and employees. Model trains and planes and coin collection provided some solace.

Mother was a passive, tense woman who spoke rapidly. She deplored the loss of openness previously existing with Carey and readily acknowledged her difficulties in obtaining his respect and obedience. Rather dramatic, she had much guilt over Carey's symptoms, feeling she had transmitted a tainted, secret mental illness since her brother had suicided in early adult life.

Also figuring prominently in Carey's life was his paternal grandmother, who frequently slipped him money to supplement his allowance and provided him with a sports car on his 16th birthday.

Born without difficulty when his mother was 23, Carey had been a fat, happy, jolly baby who learned to sing before he walked. He was trained early without difficulty. At two he showed marked hostility when his sister, Barbara, was born. He pushed her down and threw her clothing in the bathtub. His mother has vivid memories of screaming at and punishing Carey. His father expressed regrets over forcibly inhibiting Carey's aggressiveness, continually protecting him and never allowing him to fight.

In the seventh grade Carey was shifted from public school. He was lonely, anxious, and tired in the private day school, where studies required much more time. Seeing the old as good and the new as bad was a recurrent theme. Change or prospect of change was threatening. Initially hard work had earned him good grades, but with each succeeding year at the new school his average dropped. Engrossed by statistics, he managed and kept score for school athletic teams. He disdained physical education, citing his easy fatigability and the element of coercion involved. He decided that developing his physique was too much trouble and often became ill when he had to participate in physical education. He was an editor of the school paper and was widely known for his devotions to statistics, politics, and controversial editorials.

Carey and His Conservative Activities: From the initial session Carey made no bones about his intense interest in conservative politics and segregation. His minister was preaching about things that did not have anything to do with the Bible, including integration, and Carey was going to reduce his pledge to the church. His father was fearful of being ruined by liberals and for once the family was united in favor of Goldwater's candidacy. With several other students he had formed a conservative party at school, but he looked down upon some of the other members whom he considered to be atheists and nuts. His mother forced him to quit the group because of these other boys. In 1959 Carey had worked for the election of "liberals," including John F. Kennedy. This interest was short-lived, and he soon was politically oriented 180 degrees in the opposite direction.

Carey had been seen for only four sessions before the Kennedy assassination. He initially announced that the news reports of the events were boring and hard to believe, then admitted that he had almost cried, that he had felt guilty about some of the bad things he had been saying and "He was probably a pretty good fellow." As time passed, Kennedy became better and his successor became the villain.

For many months he worked avidly for Senator Goldwater's nomination and election and to a lesser extent for local "old line" Republicans. He wore hats and buttons, put on car stickers (which his father made him remove), and purchased a beverage bearing the candidate's name. He had many hopes of going to the national convention and seeing his hero nominated in person and felt quite let down when his parents would not allow him to go. He was concerned with the liberals, socialists, and communists taking over local and national government. He was convinced that if his candidate lost that there would never be another election in the country. When three civil rights workers were found missing in Mississippi he was certain they had burned their own cars and fled to Cuba—"If I got lost the President wouldn't go looking for me."

Carey staunchly maintained that he was a segregationist but not a racist. There were frequent allusions to the "colored bloc vote" which could change local and national elections. The Negroes did not know, did not understand, and appreciate issues involved, so why should they be allowed to vote? He also shared his father's views that only property owners should be allowed to vote. If his family's maid said anything out of line she would be fired, and she was not going to get a raise until a Republican was elected president. He had marked resentment when Negroes were elected to the State Legislature. He was visibly upset upon learning that the college of his choice had admitted Negro students, expressing fears that he might be given a Negro roommate—"That would be going too far!" Somewhat paradoxically he reported that his family physician had a desegregated waiting room and it really didn't bother him, for it was the doctor's personal business; besides he liked his physician. He preferred his hair long and enjoyed teenage fashions and said, "Seems like the colored people are the only ones up on style."

With much fanfare a new arch-conservative newspaper was launched in the city, its publisher being a long-time congressman who had been defeated for re-election. Carey had worked avidly in his last campaign and openly sobbed the night he conceded election—"It was the saddest time in my life." His parents stopped subscribing to the two "liberal" papers and resolved to make purchases only from merchants advertising in the new paper. At the behest of his parents Carey investigated the possibility of summer employment with the paper and was willing to work gratis, but nothing came of this endeavor.

Carey thought his father was right in not wanting him to go to an eastern college where he would be influenced by liberals and foreigners. Indeed, he had much anxiety about leaving the immediate vicinity of home, and it was only upon his parents' insistence that he consented to go to an out-of-town school. There were fears that he would know no one, and would be lonely; he half wished to fail and remain in high school for another year. In the weeks preceding the national elections his political activities accelerated. Magazines that spoke disparagingly of his candidate were hate sheets. The opposition candidates were seen as power-mad bigots and dictators, etc. On one occasion he took down the opposition's posters. He realized he could get into difficulty, but felt this act of defiance was more than justified. Motorcades and other activities provided fun and excitement. His election predictions shifted; at first his candidate would lose the popular vote but win in the Electoral College; then, "It'll be closer than anybody thinks." Upon the eve of election he seemed to be preparing to lose, comparing himself to an early Christian, saying, "In the end we are going to win." Following the election he immediately announced that his school team had won their football game, that a local candidate for whom he had worked had won his election and, "We did our part, the rest of the nation failed us." There was pride that his parents had voted against most of the proposed amendments to the state constitution. For the remaining six months of therapy his political activity was minimal, much overshadowed by concern with school, parents, anxiety about dating, and the prospects of leaving home.

Discussion

Comparison and Contrast

Although both patients wandered into the hands of a psychiatrist and were enmeshed in therapy over a considerable period of time, I should like to stress primarily the normative features presented. As Whittington[2] has noted, we have a changed tone in describing patients. Instead of writing about sick strangers, we discuss unhappy friends. In general Don and Carey had the same problems and preoccupations found in their peers. For the purpose of this paper, many aspects of their lives have been omitted in order to focus on the areas of civil rights and segregation for it was here they dissented so markedly from the majority of their classmates. Only a handful of fellow students displayed significant interest in these issues. Fewer still showed such a fervent involvement.

These two patients came from intact, relatively stable families and despite the presence of friction, breakup of the family units was never seriously considered. The fathers worked hard in responsible positions and received commensurate financial rewards; the mothers seemed to function well as homemakers. The classical oedipal triangle was evident in both families and both patients had considerable fear about sex. Don was well aware of incestuous feelings towards his mother and sisters. He dreamt and talked about them at length. He recognized his transferral of many of these feelings to the schizoid girl whom he eventually married but appeared to work out a satisfactory, meaningful relationship with her. Carey had experienced sexual feelings towards girls but in an attempt to find males with whom he could identify had obtained male exhibitionist magazines and began to wonder if he wasn't a homosexual. By the termination of therapy he was dating, wanted to kiss his dates, but had a conflict with his strict, rigid "conservative" concept of the proper thing to do.

Diagnosis: Both patients were considered to be suffering from anxiety reactions. Schizophrenia was considered in the differential diagnosis of both. Don displayed excessive fantasy and dreaming regarding sex and violence. Much of this seemed an attempt to please the therapist, his concept of what a good patient should do. Carey used rationalization, denial, and projection with marked fears of expressing directly his feelings of sex or aggression. Additionally he had a poor identification with his father. The psychologist who tested him commented, "He reflects paranoid schizophrenia."

Treatment: Both were "good patients," gave up symptoms, improved their academic performance with diminished effort and dealt with increasing success with peers and parents. Don was seen twice weekly for two and a half years and made extensive use of free association and dreams. Actually one of his defenses early in therapy was to flood sessions with accounts of dreams which he would often write down in detail. This was seen as an unconscious effort to evade the problems of day to day life, as well as his relationship with the therapist. Carey entered treatment a year after Don, was seen once weekly or less often and was terminated at the same time as Don. His treatment was more supportive and reality oriented.

Initially both patients saw me as opposed to and disapproving of their attitudes and activities. Don was sure that I would be supporting the conservative candidate in the 1964 campaign. When I informed him of my views and further let him know that very few psychiatrists would be in the Senator's camp he was genuinely surprised. Don remained committed to civil rights, but no longer had the need to provoke others by his actions. After his return from Selma he said, "Maybe I'll tell a few people and maybe I won't but I won't make a big thing of it."

After some months of therapy Carey announced that his father seemed to have improved a great deal (shades of Mark Twain!). He also voiced the previously unspoken, but obvious fear that therapy would change his viewpoints. However, "I've been telling you my view. You probably don't agree with me but you haven't argued with me." His viewpoints were unchanged; however, he could now strongly disagree with some of his old antagonists without getting headaches or stomachaches.

Role of the therapist. Whatever divergencies may exist in the background of various psychiatrists, as Ruesch[3] has observed, one element, a history of change, is common. The alteration may have ensued from childhood trauma, upward social mobility, or cultural change. It is the essence of psychotherapy that patients, by increased awareness of the meaning of behavior, become more expressive and able to adapt to life—in other words, to change. Psychiatrists have found that it is not enough to help individual patients to change. So that we may help all those who are troubled in mind, we are in the midst of trying to produce changes in elements of our social structure. Headstart, VISTA, and related programs need no introduction to this audience. I personally think such change is necessary and underscore that we, as a profession, are dedicated to change. In short, *we* are dissidents at one end of the spectrum.

Therefore, it is natural for us to study and support desegregation (e.g. 1957 GAP Report,[3] Coles'[1,4] studies, etc.). This partial involvement of psychiatrists was acknowledged by Coles[4] in relating how he came to study the Civil Rights Movement: "The standard methods of investigation, including field interviews, were simply not sufficient . . . To do meaningful research on the sit-in movement and other civil rights activity I found that another approach would be necessary—one more active, risky, and committing than any I had ever attempted. I would have to join 'the movement'." In our practices and in our research studies we see far more Dons than Careys, and I think we listen to and understand the Dons much better. Certainly one of the countertransference problems with which I struggled was to avoid overidentification with Don and rejection of Carey. The fact that I wrote up Don's history first and presented it first today speaks feor itself.

Berns,[6] in an apologia for the activists in college, romanticizes dissent as a "generally healthy aspect of the process of maturation, and many of the issues raised by the activists have vast import to mankind and our nation." He would almost seem to be saying it matters not what you change or how you change, the essential thing is that you produce change.

Youth, like riches, is most beguiling to those who no longer possess it. I have no doubt that many of us would like to be below the magic mark of thirty so that we could be trusted! Whittington[2] has wryly pointed out the enormous seductiveness of students of both sexes and the many modes of countertransference that may be mobilized in the therapist.

What Were They Striving For

It would be presumptuous to generalize from two cases. Moreover, there were no controls for factors of age, length of treatment, etc. There were no objective measurements during treatment and there was the added hazard of a therapist's retrospective falsification. Nonetheless, it was apparent that both young men were searching for identity and fidelity. As Erikson[7] says, "The evidence in young lives of the search for something and somebody to be true to is seen in the variety of pursuits more or less sanctioned by society. It is often hidden in a bewildering combination of shifting devotion and sudden perversity, sometimes more devotedly perverse, sometimes more perversedly devoted. Yet, in all youth's seeming shiftiness a seeking after some durability and change can be detected." Put another way by Friedenberg,[8] "The task is self-definition. Adolescence is a period during which a young person learns who he is and what he really feels. It is a time during which he differentiates himself from his culture, though in the culture's terms. It is the age at which by becoming a person in his own right he becomes capable of deeply felt relationships to other individuals perceived clearly as such."

In the context of this universal search for oneself I would raise the question of the relevance of the early oppositional behavior so graphically and spontaneously brought forth by both patients. Levy[9] has found generalized resistant behavior in 18 to 23 month old infants, with approximately 12 per cent of all infants showing severe resistance to examination, to being handled, etc. He has seen two-year olds with mutism, food refusal, bladder and bowel refusal, and general oppositeness. He has theorized that the capacity to resist external influence enables an organism to use and develop its own controls. He has quoted Hetzer who found that seven-year olds who had not gone through "the stubborn period" were much more dependent on teachers than others who had gone through this period. Levy viewed resistant behavior as favoring the individual's separateness and independence.

We must ask ourselves: Was the appearance of oppositional behavior in these two patients but a coincidence or is it an anlage of the adolescent-young adult dissident? Do the histories of protesters show this finding more than the population at large? If this is so, it would at least partially explain Coles'[1] finding of the relatively good mental health of his protesters and the occasions when symptoms arose when the protests were over. Does this stubborn period truly foretell later self-reliance, and mature, assertive behavior?

The direction of dissident behavior in both patients would seem to bear out the aphorism that the more people try to be unlike parents, the more they resemble them. Don's family had a tradition of changing the status quo. His mother was ostracized from church; his older sister married a non-Christian, and his younger sister had herself jailed. For a long time he viewed his actions as defiance of father and of acting out some of mother's unconscious wishes. One of the major areas worked through in treatment was his feeling that all authority figures, father, professor, therapist were out to get him and that if he happened to do anything which coincided with their wishes, he was knuckling under.

Carey's family, steeped in southern tradition, typified what Cash[10] termed "a great group of southern fears and hates which fixed itself on the line of what I have called the savage ideal . . . (to) hold rigidly to the ancient pattern, to repudiate innovation and novelty in thought and behavior, whatever came from outside. . . . As Cash[10] and Kilpatrick[11] have noted, the Southerner having known defeat in a war over slavery must, therefore, guiltily explain his views on race, whatever they are. Witness Kilpatrick's[11] scholarly defense of school segregation and former Congressman Weltner's book entitled *Southerner*,[12] his explanation of his break with southern tradition when he voted for the 1964 Civil Rights Bill.

Carey belongs foursquare anong Schiff's[13] "obedient rebels" having briefly flirted with liberalism where he found himself surrounded by radicals and socialists before returning home to conservatism where he belonged. Added to this was identification with a weak father so that Carey viewed himself as less than a complete man and had to hold on, to conserve, the little he possessed.

I have presented here vignettes from the lives of two young men struggling to find their identities and places in the world and who find themselves on opposing sides of the major struggle in American society today. I have raised the issue as to whether their tenacious involvement stems from very early life experience. Are we perhaps seeing another version of the ancient struggle where one brother fights to obtain a birthright and the other fights just as tenaciously to keep it?

REFERENCES

1. Coles, Robert: Children of Crisis: A Study of Courage and Fear. Boston, Little, Brown and Company. 1964.
2. Whittington, H. G.: Psychiatry on the College Campus. New York, International Universities Press, Inc., 1963.
3. Ruesch as quoted by Holt, R. R. and Luborsky, L.: Personality Patterns of Psychiatrists. New York, Basic Books, Inc., 1958, p. 259.
4. Psychiatric Aspects of School Desegregation. Group for the Advancement of Psychiatry, Report No. 37, May, 1957.
5. Coles, Robert: A Psychiatrist Joins "The Movement." Trans-action, Jan.-Feb. 1966, p. 21-27.
6. Berns, Robert S.: The Activist Student. Presented at American Psychiatric Association, 1957.
7. Youth: Change and Challenge. Edited by Erik Erikson, Daedalus, Vol. 91 1962.
8. Friedenberg, E. Z.: The Vanishing Adolescent. Beacon Hill, Boston, 1959.
9. Levy, David M.: Developmental and Psychodynamic Aspects of Oppositional Behavior. *In* Rado and Daniels (Eds.): Changing Concepts of Psychoanalytic Medicine. New York, Grune and Stratton, 1956.
10. Cash, W. J.: The Mind of the South. New York, Alfred A. Knopf, 1941.
11. Kilpatrick, James J.: The Southern Case for Segregation. The Crowell-Collier Press, 1962.
12. Weltner, C. L.: Southerner. New York, J. B. Lippincott Company, 1966.
13. Schiff, Lawrence F.: Dynamic Young Fogies—Rebels on the Right. Reflections, II:2-12, 1967.

Discussion by Albert Bryt, M.D.

As discussant of Dr. Cohen's poetic chapter I find myself in a position comparable to that of a cutter of precious stones who has been entrusted with a large gem and the task to find the most advantageous planes of natural cleavage. There is an inherent danger, lest the well aimed mallet land on a poorly placed chisel, leaving a totally unplanned, though much diminished end product.

Therefore, I shall refrain from an extensive critical appraisal of the two case histories so graphically portrayed by the author. Let me recall that Dr. Cohen posits the problem as follows: Both patients were searching for identity and fidelity. If I read him correctly I understand him to say that in the context of this universal, age-appropriate, search the early childhood oppositional behavior of his patients is the prototype of their search for themselves. Hence, their interest in the civil rights issue. In a more general way he seems to argue that oppositional behavior is the sine qua non of the protester. But Don's and Carey's behavior appear quite divergent to me. Does Dr. Cohen mean that all adolescent behavior feeds at the same source of youthful strivings?

Don is an alienated young man, alienated from himself and from others. He is more pervasively so than is commonly the case with adolescents seeking to find themselves. He is also rather old, at 21, to have to grapple still so extensively with finding out who he is. There is, however, a seeming characterological consistency in that his cooperative strivings are really quite rudimentary. I am thinking here especially of his relationships to his peers and fellow civil righters. Unless, of course, Dr. Cohen in his endeavor to shorten his presentation, chose not to stress the young man's capacity for intimacy.

Carey is a fearful, dependent young man. His self picture is a reflection of his parents' picture of him. He, like his parents, seems to have a tendency to make a virtue out of a necessity. He is also lonely and probably absorbed in reveries of omnipotence. I believe, also, that he is holding on to the hope that the word "no" will again reveal itself as powerful as it once was.

Neither Don or Carey in fact seriously tried to be unlike their respective parents. These young men still had to learn to care enough about themselves. At best, they wanted to be shown how to find inner peace. A striking similarity between these two patients is apparent to me in their ability to adjust in the "as if" mode. From this ability, I submit, they derived a docility which in turn made them "good patients." Neither Don, nor Carey, were dissenters; rather they were oppositionalists. Both of them seem really quite drab, even following the end of treatment. I have seen defiant youth, and I have worked intensively with a number of them. Those whom I call dissenters are psychologically alive, intellectually honest and in search of creative self realization.

Don's and Carey's life styles of bickering and faultfinding belong to a developmental epoch earlier than adolescence. They do resemble children in that David Levy's observations on oppositionalism[1] fit them better than Lawrence Schiff's picture of the "obedient rebel."[2] Both patients tried to find their selves in and by the pursuits of others. Even in this they were inconsistent. I should have preferred to describe them as: "Me too, Rebel or Reactionary, I wish I were there." The difficulties in living they met are expressive of the fact that they are but vaguely attending to the motions, rather than being actively engaged in negotiating adolescent individuation. They are still tightly enmeshed in parental identifications. Their behavior indicates a conceptual confusion between intransigeance and independence. They could not be dissenters because they were not seeking to define their own positions.

The child's oppositional behavior is a building stone of his individuation. So is the adolescent's defiant rebelliousness, but there is a remarkable difference. The child's negativistic uncooperativeness is fairly unselfconscious, and in a manner mimetic of the behavior of adults. Nonetheless, it confirms his effective differentiation out of a symbiotic parental relationship. The adolescent's dissenting self assertion, on the other hand, is very much a cognitive experience which aims at philosophic definitions of his life and of his guiding principles.

Oppositional behavior is mostly not commitment to a cause, but may be commitment against one. Dissent, contrariwise, while it bespeaks disagreement, is striving for higher ideals, a better world. It includes the prospects of risk, however, prospects of action may be secondary.

Individuation is an issue to be resolved within the individual, though by no means outside of the interpersonal context. It necessitates the dissolution of old ties, but also allows for their resumption. It is a process whereby new ties are established. In earlier years this occurred by happenstance. Now it is brought about by, and through the fulfillment of prior conative conditions. Thus individuation, together with emancipation, readiness for independence, and the ability to be interdependent, enters into the making of personal and interpersonal maturity and competence. The ability to dissent is derived from individual interpersonal maturity. However, the phenomenon of dissent-translated-into-action is a social phenomenon, contingent upon a well defined sociocultural setting, but independent of personal maturation.

Dissent, in a personal sense, is taking stock. Dissent is discovering moral values. It is stepping outside of the traditional familial and sociocultural systems, albeit to rejoin them after personal assessment. Dissent is sitting down with oneself, and working through, and consciously adopting or rejecting the principles by which one has lived. Dissent is self discovery, commitment to oneself, it is youth itself. It is, as Tolstoy has said, being struck by ideas known all along "with all the fresh force of a moral discovery," so as to become frightened of how much time has been wasted and desiring at once, "that very instant, to apply those ideas to life, firmly intending never to be untrue to them again." Tolstoy's hero says, "This moment I consider the beginning of my youth."[3]

Dissent-in-action, the social phenomenon, is a product of the American scene. When we concern ourselves with this aspect of dissent we do so as ordinary citizens, not by virtue of our professional knowledge and training. When we try to understand this translated dissent in the context of individual psychodynamics we have to go beyond our customary formulations of individual psychology. We may have to include, among others, principles derived from a familiarity with group processes. Pursuits of the group are not simply the sum total of all individual pursuits.[4] We may have to avail ourselves of insights offered by the sociologist. Consider this observation by Parsons that "youth has become a kind of 'loyal opposition' to the main trends of culture." Recently, he says, there seems to have been a resurgence of political interest and activity. It has tended to focus on specific issues in which moral problems are sharply defined, notably race relations and the problems of the nuclear war.[5] Had Parsons written this paper in 1967 instead of in 1961 he would, undoubtedly, have added the war in Vietnam as a meaningful issue. Indeed, Parsons makes the point that youth is concerned with the meaning of actions and events.

The person who is self-absorbed has no leisure to devote himself to other meaningful issues. Moreover, unable to share with others their concerns, he cannot share his own concerns either. Thus, he is left with an increased sense of isolation. In essence this was Don's dilemma during and after Selma, "the symbolic masturbation in front of the big State Capitol."

As practicing psychoanalysts we are probably more concerned with, and value more highly the meaning of what a person does, than might other segments of the "well established" community. Hence our sympathies with the youthful strivings of dissenters-in-action. However, when it comes to the role of dissent-in-action we have no other claim to qualification but that of individual citizens. Our specialized training in psychoanalysis does not equip us for such expertise. Contrariwise, it is our position as professionals, very specifically, which enables us to offer useful assistance to those who call on us for help in their quest to find themselves.

References

1. Levy, David M.: Developmental and Psychodynamic Aspects of Oppositional Behavior. In: Rado and Daniels (Eds.): Changing Concepts of Psychoanalytic Medicine. New York, Grune & Stratton, 1956.
2. Schiff, Lawrence: Dynamic Young Fogies—Rebels on the Right. Reprinted from Transactions in Reflections, Vol. II, No. 2, 1967, pp. 2-12.
3. Tolstoy, Leo: Childhood, Boyhood and Youth. Translated by Michael Scammel. New York, McGraw-Hill Paperback Edition, 1965.
4. Hare, Borgotta and Bates (Eds.): Small Groups, Studies in Social Interaction. New York, Alfred Knopf, 1955.
5. Parsons, Talcott: Youth in the Context of American Society. In Youth: Change and Challenge. Daedalus, Winter 1962, pp. 97-123.

THE PSYCHODYNAMICS OF DISSENT

A Clinical Appraisal with Emphasis on Racial Activists

By CHARLES A. PINDERHUGHES, M.D.

The General Meaning of Dissent

DISSENT IS A TOOL for updating the life style of a culture: the culture of a family, of a group, of a society. The essence of dissent is opposition expressed from a position which is considered inferior in power, in numbers, or in substance. In every group expression of dissent, a wide range of individual meanings and motivations can be found. In every individual manifesting dissent consciously, a variety of latent and unconscious meanings and motivations exist.

The psychology of a dissenter has meanings derived from the groups to which he relates. A true dissenter is motivated primarily by feelings activated by the issue. A pseudodissenter employs dissent primarily as a means to an end which lies outside of the issue. By dissent he achieves a longed-for role or relationship, or some other gain which heightens his sense of value. True dissenters manifest varying degrees of pseudodissent, especially when external threats mount or their ego resources become depleted. Under such circumstances dissenting posture may be maintained and motivated primarily through support of associates.

Since propriety of behavior, sanity and intelligence generally are defined in terms of conformity at a community level, those who dissent with vigor in the midst of stable social orders are quite apt to be considered as inappropriate, insensitive trouble makers who are "crazy," "kooky," or at least "stupid." The relative isolation, disapproval, and massive resistance they experience means that they require a wealth of internal personality resource to support the dissenting position.

From the Department of Psychiatry, Tufts University School of Medicine and the Psychiatry Service, Veterans Administration Hospital, Boston, Mass.

In an age of dissent, fewer internal determinants are required and a wider variety of personalities can take a dissenting position in the presence of acceptance, support, or reward for doing so. In any social order the potential for stability or for flexibility and change may be related to the ease with which dissenting groups may be mobilized.[1]

During periods of relatively stable social structure, intense social and psychological pressures induce conformity in all but the most independent, the most courageous, the most alienated, or the most conflict-seeking persons. Changes occur slowly and against high resistance in such social orders.

In periods of reduced stability when change in the life style of a culture can be brought about more easily, more individuals express dissent. As the ranks of dissenters grow, new leaders, new identifications, and new directions for dissent evolve and fan out in many directions. Common motivations, common sentiments, common methods, or common targets unite dissenters into groups of varying size and purpose, and dissent becomes a basis for conformity in these groups.

Conflict, anxiety, and activity are at a high level and changes in life style and new group formation are common. Persons threatened by the conflict, anxiety, activity, and change are mobilized toward counter-activity to restabilize the social order.

The foregoing general statements about dissent offer a basis for

Summary of Cases Showing Degrees of Dissent

	No Social Dissent	Mild Social Dissent	Strong Social Dissent
White			
1952 - 62	29 (76%)	4 (10%)	5 (13%)
1962 - 67	30 (50%)	11 (18%)	18 (30%)
Negro			
1952 - 62	9 (75%)	0 ( )	3 (25%)
1962 - 67	4 (23%)	4 (23%)	9 (53%)
Total			
1952 - 62	38 (76%)	4 (8%)	8 (16%)
1962 - 67	34 (44%)	15 (20%)	27 (35%)

Total Cases: 1952 - 62 50
1962 - 67 76

hypotheses which may be evaluated by examining clinical records of patients in psychotherapy and psychoanalysis.

Hypothesis I

When a social order is relatively stable and dissent is "endemic," the number of patients actively engaged in dissenting social action is apt to be small and the psychodynamics of those patients relatively circumscribed.

Hypothesis II

In an age when dissent exists in epidemic proportions, the number of patients actively engaged in social action is apt to be large and the psychodynamics of those patients quite varied. The only characteristic personalities which may not be found among dissenters are those unable to tolerate change or closeness. Such patients are apt to be chronic dissenters against change or against intrusion.[2]

Method of Study

Part I: Clinical records of all private psychotherapy and psychoanalytic patients treated by the author in the fifteen-year period between June 1952 and June 1967 were reviewed. Patients who found dissent on a social issue to be emotionally meaningful and lent vocal, written, financial, or personal support to dissenting social action were listed as dissenters and arbitrarily divided into mild and strong groups.

The psychodynamic pattern for each of the strong dissenters was reviewed and brief descriptions recorded to demonstrate characteristic features of several strong dissenters. All patients who showed no dissent on a social issue were reviewed for comparison with dissenters.

Part II: Prevailing themes in clinical records of dissenters were examined, and significant themes in nine patients and one nonpatient were described.

Part III: Excerpts from several interviews were selected to portray some dynamics more graphically.

Part IV: Clinical observations of social interactions in two hospital settings were studied to determine the impact of social factors upon the psychodynamics of dissent.

The meaning of the data is discussed and an attempt made to develop general implications of clinical or theoretical significance.

Part I. Characteristics of Dissenters Treated 1952 to 1962

The cases referred to in this section are:

a. The four patients (Cases 4, 5, 6, 8) treated between 1952 and 1962 who are known to have manifested dissent on a social issue between 1952 and 1962.

b. Four patients (Cases 1, 2, 3, 7) treated between 1952 and 1962 took no dissenting action on a social issue prior to 1962, but were observed to do so afterward. (The absence of additional follow-up information on patients treated between 1952-1962 makes it likely that some of these who took dissenting action only after 1962 have been omitted.)

Brief profiles of the eight dissenters treated between 1952 and 1962 portray the similarities and differences in these patients. Libidinized conflict-full relationships were common, but otherwise differences between the patients were great.

Case 1: This patient was reared in a large black ghetto by foster parents after the death of his parents in his second year. He was treated for inability to finish a task, a sense of imminent disaster, self-defeating erratic behavior, and suicidal thoughts. He practised brinksmanship with his health and behavior, sought direction and control but openly fought against those who gave it. Obsessional and depressive symptoms were present in his passive aggressive personality.

Case 2: Reared in a middle-class family in a large black ghetto this patient had an hysteric character with some obsessive and depressive features. A breakthrough of repressed sexual wishes and breakdown of reaction formations led to marked guilt feelings, constriction of behavior with withdrawal, and loneliness. She identified with and took action on behalf of people who were "hard to reach."

Case 3: Reared in a middle-class family in a large black ghetto, this patient has a passive dependent personality and repeatedly developed conflict-filled attachments with alternation of roles in which he experienced disappointment and rejection. He began treatment when his wife left him. His principal social action was taken on behalf of neglected children.

Case 4: Reared in a middle-class family in a large white ghetto, this patient was mildly depressed and obsessional with affect isolated. In a quiet committed way he was intensely motivated to social action designed to assist distant helpless people to communicate better and to fend for themselves.

Cases 5: Reared in a wealthy white family, this patient entered therapy with complaints of "vagueness," inability to control behavior, and "disorganization" which occurred every two to three years. He had a borderline disorder, periodic schizophrenic symptoms, and considerable sexual disturbance. Motivated on one hand by a sense of being a wild, untamed maverick and on another by being a privileged aristocrat, he travelled extensively and delighted in developing images of himself as a remarkable executive who made large numbers of people "dance to his tune." Living out of a controversial, attention-getting role similar to his "spoiled child" role of early childhood was prominent and his dissent on a social issue incidental and transient.

Case 6: Reared in a middle-class family in a large white ghetto, this patient was an impulsive, infantile, hysteric, young woman who viewed herself as unloved, uncontrolled, fat, messy, and rebellious. Her behavior provoked an endless series of intense stormy brief relationships in her personal life.

Case 7: Reared in a white middle-class family in a small town, this patient was an hysteric young woman with conversion symptoms, depression, obsessive compulsive and borderline features. Early bowel problems, enemas and struggles with her mother were antecedents to her negativistic opposition to her mother. A rebel without cause during the middle 1950's, she developed interest in social action only after considerable improvement in 1962.

Case 8: Reared in a well-to-do white family in a suburban town, this patient had hysteric symptoms with depression. She felt "low, under a yoke, inadequate, unequal, and rejected." She had a stormy marriage and alternated with her husband between roles of a troubled forlorn messy child and a disciplined rescuer who put things in order.

Discussion of Dissenters Treated 1952 to 1962

Of the eight patients treated between 1952 and 1962 who were listed as strong dissenters, only four (Cases 4, 5, 6, 8) showed any dissenting social action prior to 1962. Thus only four cases out of 50 (8 per cent) showed dissenting social action when dissent was "endemic." They represent too small a number for any conclusions to be drawn. However, it is notable that they show such variety in dynamics that no characteristic circumscribed profile could be said to exist even for this small number. Three (Cases 4, 5, 8) of the four went outside of the United States to work with distant underprivileged people while the fourth (Case 6) worked for an agency whose principal purpose was social protest. It is also notable that none of the three Negroes in this group showed strong social dissent prior to 1962. It would appear that between 1952 and 1962 strong expression of dissent was sufficiently unpopular to discourage many persons who were strongly inclined in that direction.

During this period only one of the patients expressed dissent as an individual acting independently (Case 5). Three others expressed dissent in their work with large highly respected and well-known organizations. Two of these involved a missionary motif in that they were in foreign lands (Cases 4, 8).

The one patient expressing dissent as an individual, wihout support of other persons, had very disturbed interpersonal relationships and was the only psychotic person among the eight dissenters. Among the 38 nondissenters were five psychotic and seven borderline patients.

Obsessional patterns or conflicts associated with bowel functioning seemed more common in the clinical records of these patients than any other syndrome, but this was characteristic of the nondissenting patients also. This may have been partly determined by countertransference elements in the therapist. In patients dissenting on social issues or only in family or work situations, struggles over discipline, over who had control, with much libidinized opposition and withholding, or with precipitous impulsive discharge, seemed to be present frequently in persons with strong inclinations to "take a stand," to oppose, and to go on strike, or otherwise demonstrate.

Characteristics of Dissenters Treated 1962 to 1967

The review of patients treated over a fifteen-year period disclosed the sobering fact that every patient at some time during treatment expressed dissent in some family or social group context. Material from patients who did not show dissent on social issues revealed no characteristic differences from the group which dissented on social issues. In fact, both groups were full of dissent but varied the targets toward which it was directed. Many dissenting personalities brought forth no dissent on social issues but vigorously dissented with relatives, teachers, employers and others.

While the age, sex, and sources of the patients have remained relatively stable the number of patients expressing dissent on a social issue in some manifest act has greatly increased. From 1952 to 1962 few patients found dissent on a social issue to be emotionally meaningful. Between 1962 and 1967 there developed a rapid increase in this number to a point where social issues ranked high among arenas chosen for expression of strong feelings and beliefs of a dissenting nature. Between 1952 and 1962 one patient in every four actively and publicly expressed dissent on some social issue. The frequency was the same for white and Negro patients. From 1962 to 1967 two of every four white patients and three of every four Negro patients actively and publicly expressed dissent on some social issue.

Although 35 dissenters were reviewed, reference is made in this chapter to clinical material from only 20 patients. Six were Negro males, four Negro females, four white males and six white females. Included in this number are one organizer of a chapter of a civil rights group, five persons who were quite active in civil rights groups and who resigned from prior jobs in order to work on black ghetto problems, one person engaged in research efforts supporting civil rights activity, five persons who worked in underdeveloped countries overseas including one in Africa, two in the Far East and two in the Middle East, two employed by organized social protest groups, one nationally known racial activist from the South, one peace and anti-Vietnam War marcher, two previously investigated ex-communists (one with family members jailed), two former Muslims, one follower of Malcolm X who was present at his assassination, one SNCC worker in the deep South, and one early Freedom Rider jailed in the first Mississippi Freedom Rides. All of the Negro dissenters supported Black Power concepts of one kind or another.

Part II. Selected Significant Themes in Dissenters

Case 9—Significant Theme: This patient lost a dominating mother at an early age. He identified with her by taking over housework and activities related to the mother's occupation. He experienced little mourning following a series of losses, was hypomanic at times, often in grave conflict with family or with others, fearing loss of control. He clearly searched for a strong controlling father figure, the father having died just before conflict in the patient's relationships escalated to a painful degree. Attempts to contact, manipulate and control parental figures, while depending upon them, characterized his behavior in family, social, and civil rights spheres.

Case 10—Significant Theme: Although involved in many conflicts with social agencies and with her family, this patient had a need to find and devote most of her time to helping people who were needy. Hysteric features were prominent with bisexual fantasies lived out in clothing, behavior, and performance. At times she felt she had a penis, at times not. She alternated between pants and dresses. She reguarly made A's in languages and failed in sciences. Disappointed with her passive father, she fought with her active mother whom she saw as "on her back" pushing her to do better. She maintained membership in dual groups of opposing values which she spoke of as "a typical crowd" and of "a way-out beat crowd."

Case 11—Significant Theme: This patient searched for intense attachments while wishing to be chased and beaten by a phallic figure and rescued by a nurturing one. Subsequently she provoked the nurturing figure to be an attacking one by her uncontrolled messy behavior. She felt undeserving and unloved, periodically perceived herself as black, and engaged in much self-defeating behavior. Civil rights activity was undertaken clearly on behalf of that part of her personality which was symbolized by "Negroes." Rescue of a fearful excited figure fleeing from sadistic attackers was the fantasy motivating several years of devoted work with a civil rights organization which itself was viewed as immature, unstable, weak and under attack.

Case 12—Significant Theme: Obsessed by being "greedy, avaricious, hungry for food, sensory stimulation, and knowledge" of all kinds, this patient made demands of those close to her which were perceived as overwhelming, dominating and destructive. Extremely sensitive to "being pushed around," she engaged in fierce struggles with powerful figures to keep from being dominated and to maintain control. Importantly, such behavior kept her in vigorous interaction with authority figures. She had been a brilliant but stubborn child, developed a capacity to master and retain feces without a bowel movement for many days in willful, stubborn struggles with her mother and with doctors. She repeatedly developed alliances with seemingly just but unpopular causes and organizations for many years.

Case 13—Significant Theme: This isolated, obsessive compulsive white male felt he had been left by his parents in a field to wander. He viewed himself as a lonely child without adults, searching for food and human contact. He felt his parents had "gotten rid of him," and he formed loving relationships with animals on the farm. Obsessed with fantasies of setting fires as a small boy, he later planned to be a fireman. He engaged in solitary social protest, interracial marriage and self-sacrificing service to backward people.

Cases 14 and 15—Significant Theme: Two male patients engaged in vigorous social protests in ways which interfered with their work as professionals. Both were from large families which lost fathers. Both felt plagued by ever-present nagging mothers and wives. Both had fantasies of a penis in the rectum (one feared being a freak like the bearded lady, while the other fantasied pregnancies in the rectum delivered by defecation). Both were conscious of intense love for the therapist. Both periodically had paranoid symptoms which could be relieved quickly by interpretation of the strong positive transference. Both cried following such interpretations. Both got into such difficulties in their work that the therapist drafted letters for each which prevented them from being fired. Wishes to have fathers behind them, giving them motivation, discipline, training, and a penis were prominent in both. Both carried out their social protests within the framework of their jobs and thereby brought severe criticism upon themselves.

Cases 16 and 17—Significant Theme: Two patients enjoyed civil rights work in the South where they felt themselves to be in an adult warmly helpful role, more active and more mobile. They felt as if they had abilities, personal equipment, and stature which they did not have in the North. Some of this sense of importance related to their image of the South and of Southerners as less adequate, less competent, more needy (i.e. more childlike by comparison). Yet in both instances the content of interviews regularly revealed relationships with mother figures to be important to the sense of warmth and adequacy in the South. One of these patients was white, one Negro, and both felt guilty for going South and feeling superior.

Case 18—Significant Theme in a Nonpatient: As an official representative and spokesman for a black Muslim organization, this man came into vigorous debates with non-Muslim blacks and whites. An intelligent man who tested reality well, he conveyed understandings which gained many supporters for his Muslim group, but in turn he identified with an increasing number of non-Muslims. As his faith weakened, he first defended his group staunchly and then left the Muslims to assume a more militant position and to advocate violence. A sense of independence accompanied this phase as he moved from the Muslims to a loosely organized militant group and then to an independent position. He has since worked capably in community agencies against which he previously fought. He sees Negro self development, self segregation, and militance as a stage preparing Negroes for integration. For him racial activism appeared as a stage in his personal growth and development. (Many Negro civil rights workers have described feelings of maturation of their own personalities during their interpersonal involvements as the principal gains the movement has experienced).

Part III. Selected Excerpts From Interviews of Dissenters

Case 19: This patient and several others showed an increase in the intensity of civil rights activity prior to and during interruptions of treatment and at other times of heightened conflict in the transference. The following sequence of ideas was presented in one interview as this intelligent aggressive woman used social action as a defense against the positive transference she perceived as weakening her. "I'm leaving early—I have to deal with a man at an important meeting—he has screwed everybody except me—I've got to hold my end up or he'll screw me too—that confrontation between rich and poor is the same as between black and white, and I feel obligated to play a role in the process—nice liberal middle-class whities see me as close to their side. They show me favor and try to win me over, then the poor and the black people can be manipulated. I'm black. I represent the worst in people. (At this point the patient cursed and was asked by the therapist how she felt about cursing). It gets my point across. I use it a lot. Women don't do it. I don't now use it although in my neighborhood it's done a lot—I've got to go (she started to take a pill and was asked what it was). It's antihistamine. I suddenly started to get hayfever (the therapist recommended that she not take the pill until he found out what she meant by hay fever). It means itching throat, running nose, and running eyes. There's something in the air from your air conditioner—I developed it yesterday, too, just as I was leaving a wonderful old lady who is very motherly to me." (She repeatedly asked to take the pill but was advised not to take it although it was acknowledged that she herself must decide what to do. She took the pill, blew her nose, wiped her running eyes and arose saying that she was sorry she had to leave. The therapist interpreted that she wished to cry as she leaves motherly people, that feeling wells up in her just as she leaves them. Her going out to dissent against the man who is out to screw her was interpreted as acting out. She was told that she felt she must fight against the therapist to keep from feeling weakened, vulnerable, hurt, overwhelmed and screwed as she begins to feel more like a woman. She was told she defended against this by going out and fighting like a man.)

Later on in the treatment of Case 19 as considerable improvement was manifest, dissenting actions continued but were undertaken with more judgment and less emotion. Consider the following quote from the same patient: "My indignation grows. Glass on the street caused the flat tire which made me late. Periodically I got livid, furious and helpless. Like finding poor quality vegetables and prices elevated on pay days which are well-known. I've lost interest in doing things I usually do with fervor. It's just not there. If there's a job to be done I used to be the hardest working person around. Am I becoming a very inane person? My distinguishing characteristics have been slowly stripped away as I get my personality conflicts resolved. I don't lose my temper, don't do things on impulse and I'm not terribly enthusiastic about anything. Things that made me mad are just not there any more."

This patient also demonstrated an important feature noted in the behavior of many persons engaged in civil rights activities with a small tightly knit group of high morale. She gave completely of herself, getting inadequate sleep and inadequate nourishment, and overworking until exhausted or sick. In early phases of her treatment she could not tolerate any thought of functioning in another way, and her view of other persons' needs and the importance of her response was such that the therapist compared the reaction to the feelings of a mother toward the demands of her children. In fact this particular patient disclosed the fantasy of being the mother for all the Negro and poor people in an entire city.

Case 20: The following sequence of ideas appeared on one interview of a patient whose social protest was based on feelings of mistreatment by all father figures and all men. "First I forgot where the check was and then I did not sign it. I can't remember now. I feel lost, insecure, unable to hold onto anything. I'm caught between giving and withholding; I fear I'll be taken advantage of so I play dumb and get people to lash out at me and then I fail socially. I'm a griper. I turn against myself when I hate others' guts. No decent shrink would let me do what I've done. Dr. X was a stupid asshole. He let me shit on myself. He took the finger out of the dike and I had to contend with the flood. Years of talk to stupid people. So I don't give in, I don't produce. I'm quiet because I'm less of an ass that way. That's why I'm so passive."

From Case 16: The following sequence of ideas were presented in one interview after the patient shifted from wearing pants regularly to dresses. "I don't want to think about a bloody rag I saw—I wonder about menstruation or about someone being cut—I recall encountering a man who lost his legs—

things were messy last week with my husband; I was let down and he made me feel bad. I think I have lost him—I ws ambivalent at Stokeley's rally—I cried when old Mr. Jones called—I thought of you—of losing my father when I was four years old and how I overreact to losses—I remember my yellow dress, my first menstrual period. My mother had not told me. I felt it would stop if I would not notice it. I was told by a boy that I had paint on my dress. I like to paint. I was considered a brown nose—a good house nigger who cleaned all the time. I recall a dream about my nose and a cup without a handle. (The discussion with the therapist focussed upon the way in which her sense of losing her husband reactivates the loss, the feelings of losing the father at age four and her sense of losing her body integrity after fantasying herself as a boy. She was told she seems driven to repeat the loss in various ways and that she tries to ignore what she is doing. Her engineering of the loss of her husband, her denial of this, and her failure to clean things up because she relives the old fantasy of being a bad, messy, uncontrolled little girl who loses father and is hurt were emphasized). I feel he is only mine for a little while and that I won't be able to keep him."

From Case 12: An object-seeking patient reacted with rage in protest against the therapist closing his eyes in thought. In one interview the following sequence of ideas was expressed: "You looked vigorous yesterday. After my vicious attack you had something taken from you and were downcast. I felt better then but I felt guilty. I separate myself. I hated to feel shut out and away from the action. I couldn't tolerate my mother talking on the phone when I was small or people talking in another room. There seemed to be big mysteries that I was not in on. I'm hungry and go hog wild, feeling excitement, thoughts of kicking you and throwing things at you."

From Case 20: A previously inhibited, obsessive male entertainer produced the following sequence of ideas in an interview: "I don't feel I deserve good things or pleasure—entertainers are exploited and with my position in the Union I need to look after them—many of them seek to be free of commitments, alliances, and time-consuming encumbrances and this helps them to be exploited. I should protest the exploitative treatment directly in a massive way immediately. (The therapist advised that he not do so now because destructive feelings just emerging were not yet sufficiently mastered and attempts in this direction might sabotage his efforts to organize his life more soundly). I reject that. It's a compromise. It's giving in to the existing system which is terrible to buck. If I send off a long protest letter I'll feel better to get it off my chest, but retaliation would be damaging to me. I fear I'm small-time, and yet many look to me for leadership. I fail them because I fail myself. It was impossible to please my mother. I couldn't get praise, felt I never deserved anything. Now I'm embarrassed to be praised and can't accept compliments. I'm only comfortable when I take a low role, a back seat and complain. I feel rage at all the mistreatment, want to hit the first available person. I did. I hit a poor college kid last weekend when I should have hit the manager. I'm so afraid of getting out of control. I suppress it."

Case 21: In this early Mississippi Freedom Rider, memories of trauma experienced during dissenting social action were linked to repressed content from early emotional trauma and frequently heralded threatening developments in the transference. Frightening psychophysiologic reactions occurred during termination which had been experienced previously only when jailed and filled with fear for several weeks as medical attention was denied her. At an earlier point in treatment she presented the following sequence of ideas just prior to a two week vacation.

"Did you push me away purposely—I feel strange, separated from you, lost, adrift—frightening dreams in which I am menaced . . . I awakened terrified—I remember when I was taken from the demonstration and arrested—there were no friends, I didn't trust anyone, much was in the newspapers including pictures of the crowd surrounding me—people yelled 'lynch them.' I felt surrounded by latent violence, alone and deserted—like the time I explored a cave—or went mountain climbing to overcome fear of heights—there's a primitive feeling following streams in a cool dark cave—I have obvious fascination in being frightened and impulsively took trips—someone would say: so you want to explore a cave, okay, lets, it's lots of fun—I'm talking about my vagina and have feeling there—I feel weak, as if I look down from a height—when I practice rapel—one person at top plays out a rope as the other goes down a steep cliff. [Her faith in the person at the top was discussed and her anxiety related to the impending separation.] I do wish to get back to my mother. When I was arrested she rejected me, pushed me away. Sometimes I do feel like a motherless child."

Part IV. Clinical Observations on the Impact of Social Factors Upon Psychodynamics of Dissent

Example 1

Upon examining the basis for widespread dissent in a mental hospital the author found the following. Several years before there had been a well-trained intelligent clinical director whose style of administrating was deeply entrenched. Staff members with embryonic ideas took them to the clinical director who assisted them in the development of the ideas in a vigorous interaction with them. He was loved, respected, and when he left the hospital there was a strong sense of loss. The replacement was also well-trained and intelligent but his style of carrying out his work was quite different. Instead of a competitive approach in which a spirited struggle was enjoyed, defensive patterns were more prominent, and he was inclined to examine each suggestion brought to him in the context of existing rules without offering stimulation and assistance. He was inclined toward conservatism rather than change. The replacement experienced considerable dissent on part of the hospital staff which was based partly on the sense of loss and bitterness. The staff also expressed dissent against the replacement for not meeting their needs, the pattern of which had been conditioned by long experience with the first clinical director.

Example 2

Several years ago the author studied the effects upon patients and personnel of loss of both ward doctor and head nurse on a ward. The effects of the double loss were attenuated by introducing replacements who made no decisions, gave no orders, and provided no structure. Persons with inadequate internal controls began to act in less controlled fashion which activated dissent by other patients and personnel. Excitement in patients and personnel increased and a disturbed social state with many fires and other crises developed. Crises regularly gained attention and care and invited control. Megalomanic patients could not be controlled consistently for many weeks, and they produced a lengthy series of crises which induced longed for, vigorous struggles with the personnel. Dissent, social disorder, and change in social structure continued at a high level for four months during which time intense relationships, group formation, and high morale developed in a setting of considerable action and excitement.

At the same time, widespread dissent within a social group can express feelings of loss of desired parental services, a wish for those parental services, and a provocative act designed to produce the desired services from figures on the scene. The individuals involved in Examples 1 and 2 were quite unaware of the significance of the factors described. Massive dissent was focussed upon the clinical director in Example 1 until he was removed. The patients and personnel involved in Example 2 felt themselves to be in the midst of chaotic riot-like circumstances until they were able to figure out (in many tape-recorded meetings) the dynamics of the social disturbance on the ward and how they could successfully manage it. In both examples dissent prevailed in epidemic proportions until an authority figure with desired characteristics was reproduced.

Discussion

Clinical Implications

The clinical material studied did not support the hypothesis that the psychodynamics of dissenters are relatively circumscribed when their numbers are small, or that personality disturbances were significantly deeper or more extensive in dissenters than in nondissenters on social issues. The clinical material disclosed that patients dissenting on such issues showed a wide variety of personality structure and clinical findings. It further disclosed that all patients directed dissent toward various targets in the course of their lives. Patients in whom dissent was prominent tended to have vigorous libidinized conflictful relationships whether they choose to express dissent on a social issue or in some other area.

What, then, can be considered appropriate professional activity for the psychoanalyst who encounters in a patient dissenting action on a social sphere? The dissent should be welcomed as an affect laden presentation to be analyzed and understood for what it may contribute to treatment. Analysis of reactions to the issue and of relationships associated with the dissenting posture is invariably fruitful and may constitute the bulk of the early therapeutic work with dissenting personalities for whom the issues and persons embedded in them represent the most libidinized relationships. When therapeutic efforts are successful, the immature elements of behavior will be brought under management of the mature ego and dissenting activity will be reality oriented and without a driven quality.

For those interested in psychoanalysis or psychoanalytically oriented psychotherapy, these findings carry, for the most part, messages relative to transference and resistance in the treatment of persons who manifest social acts of dissent. The sensitivity of dissenters to the world around them makes them perceptive of and responsive to the behavior of the analyst. Attention to the countertransference may be especially important in the treatment of patients with strong needs to provoke authority figures into opposition posture. I have found no special considerations or deviation from standard technique to be needed.

The findings in this and other studies may be coherently synthesized around the scheme employed by Erikson[3] for delineating interpersonal issues encountered in psycho-sexual development.

Discussion of Childhood Origins of Feelings for Human Rights

For every individual the first human rights crisis occurs at the moment of birth. The loss of such comfort, care, and security for which one had developed a constant expectancy is shocking. Infants generally respond with an immediate protest, and subsequently demonstrate regularly to ensure their rights and entitlements while they attempt to recreate symbolically the Garden of Eden from which they were cast.

From the beginning infants are actively trying to dominate, control, and enslave mothers.[4] For the young infant a good parent is a good slave. A second major issue over rights is often precipitated when mothers compete with their babies for consideration and control.

The mother starts a third conflict over rights when she weans the infant, and a fourth by introducing early behavior training. The "slave-mother," by helping to develop feedback and self-governing mechanisms in the "master-child" remarkably alters the relationship so that she can make demands upon, as well as accede to, her child. A large increase in potentials for mutuality in relationship occurs at this point.

The child's development of some mastery and discipline in using complex internal psychomotor patterns is an essential prerequisite and important tool for the later struggle to master the external world.

In the anal and early phallic periods, sufficient neural development has occurred (between 9 months and 3 years) to make possible the establishment of some self control, competence, independence, and an internal sense of authority. A well known period of "do it myself" negativism (which might be thought of as a development of Child Power) is employed to modify both the early identity of the child and the role relationships with those on whom the child has been dependent. This phase of relationship finds the parent and child sometimes cooperating, sometimes opposing one another, and power struggles with issues of right or wrong are common. While parental power may appear to prevail on immediate issues, child reactions in behavior, symptoms, or maladaptive character traits subsequently may indicate that the process or outcome was unsound. With internal impulses and apparatus mastered, a fifth conflict takes place in the phallic struggles to master the external world and to determine who stands firm and who surrenders.

The sixth and final group of childhood crises over rights have to do with conflicts in ideology. The possibilities for conflicts in values, principles, and ideas provide rich and endlessly varied battlegrounds for those inclined to fight for the preservation or extension of ideas with which they identify. The rights to integrity of beliefs can seem as important as rights to integrity of body.

By the time children are six years old the parents not only have been emancipated, but have taken over the slave driver's seat.

Hopelessly overwhelmed, the children bide their time in a latent period without wars until they gain in strength of mind and body and band together to organize the adolescent wars, for independence from one group accompanied by dependence upon another.

They struggle to free their minds of bondage to their training, to expand their consciousness, and to escape from old conformities into new ones. In these adolescent conflicts, the rights to rebel, grow, and develop power and identity are prominent once again. The gap between youths and adults in physical, mental, and emotional resources closes progressively. These resources become organized into an identifiable pattern with the assistance of youth groups and varying degrees of cooperation and resistance from adults. Each youth works out his pattern of intrapersonal and interpersonal psychology with group members of his own kind (in sex, interests, cultural features, etc.) until persons of different sex, interests, and capabilities offer possibilities of a more gratifying complementary pair of role relationships.

Gradually, one by one, the adolescent gang breaks up, puts down the idealistic sword and shield, establishes conventional adult relationships, joins conventional adult establishments, experiences disappointment and expresses dissent in conventional adult ways, and brings forth a new generation to repeat the process.

Extension of Rights Conflict to Other Arenas

Each child's primary struggles with its parents over rights are relived, reinforced, and modified in the arena of other family relationships.

With widening of the scope of interactions, conflicts involving rights are extended and transferred in succession to every new relationship, and to every new group in which one participates. Struggles between student and college administration, between employee and employer, and between marital partners, commonly involve conflicts over rights.

Sooner or later all individuals encounter conflicts with civil servants, police, laws, and government officials at which time they experience highly personalized feelings on civil rights and become, in a sense, civil rights workers. When such feelings are acted upon in concert with enough additional persons a civil rights movement may develop.

Child-Parent Conflict Over Rights as Prototype of Civil Rights Conflict

While separated by a complex metamorphosis it is possible to see that the infant-mother conflict over rights serves as a prototype of civil rights conflict. It is well to remember that the infant may behave quite reasonably and that the mother may behave quite irrationally. Some mothers are regularly more infantile and less reasonable than their babies. All mothers at times and in some areas show infantile and unreasonable behavior. So it is with social groups and with governments.

The transfer of feelings from childhood to subsequent experience virtually ensures that some persons will continue seeking to control, to exploit, and to destroy others, and will use every available social organization to further these goals. There will continue to be ongoing, dynamic struggles between those seeking to gain or preserve opportunities for exploitation and those seeking to restrict such opportunities.

Some basic protests are shared by all individuals and groups who feel that their rights are limited unjustly. Children, fatigued mothers, adolescents, abused wives, and henpecked husbands, women before suffrage, and laborers have shared many of the protests. The statements of protest may be arranged in a sequence in which they might appear in personality development.

"You can't eject us like this, after carrying us so long."

"You don't trust us."

"We don't receive the care and nurture we need. You reject and neglect us. You aren't giving, considerate, or fair. We are not properly weaned."

"You have made us devoid of authority necessary for self-discipline and internal control. You won't let us be free."

"You think you are mature and you so rationalize your actions that you can't see how you are babies omnipotently demanding and controlling feeding, cleaning, and work services from us."

"You always exclude us, or place us at your back, or your bottom, and treat us like feces. You encourage us to sacrifice and to defeat ourselves. You force us into inferior roles and have constructed a society and government which ensures it."

"You have rights to integrity of your body but we have no such protection. You exploit and brutalize."

"You brainwash us with your ideology of your supremacy and of our inferiority. You won't even listen to and seek to understand our ideology especially when it fits the facts."

"You can assert yourself but we are not permitted to do so."
"You hold up guiding principles and values to rationalize your humanity but you don't follow them."
"Your backlash again beats our backs and we must bide our time and gain in strength and power in a latent period."
"You refuse us peer status, refuse to permit us into your groups and your establishments."
"You drain off our leadership one by one into your establishments which gradually neutralize and brainwash them to fit the patterns of the establishment."

All dissenters show one of the above protests or some variant thereof. Moreover, each dissenter employs a pattern of dissent derived from the ego operations commonly employed to manage that protest in childhood.

The childhood experiences provide the sensitization and capacity to react to rights issues, and define the preferred pattern and the intensity with which reaction occurs.

Regardless of the direction of the main thrust of a rights movement, the wide spectrum of childhood experiences reactivated provides a fanning effect which will widen the scope of the movement and enable new developments to unfold.

It is notable that the sequence of stages which emerged in the Civil Rights Movement roughly paralleled the sequence of stages in personality development from dependency, to passive-aggressiveness, to active mastery. The nonpatient in Case 18 shifted over a four year period from negativism to passive-aggressiveness to verbal dissent to militant dissent to independence to identification with those dissented against. He underwent this remarkable metamorphosis while continuing to work toward beneficial social change. Most dissenters are not static, but are undergoing changes in methods and behavior under the impact of their experiences. It seems possible that, in paired parties with disparate power and one dependent party, whenever the dependent party seeks peer status against the resistance of the stronger party, the process proceeds through a sequence of stages comparable to those of personality development.

Anti-Dissent Prejudice and Discrimination

The dissenter usually faces superior power and authority which devalues and attempts to discredit his dissent by various means.

Most persons visualize dissent occurring from an inferior position and rising against resistance to threaten higher levels. The disparate values and prejudices associated with upper and lower positions virtually guarantee that lowly dissenters will generally be discriminated against while their targets are glorified. The prejudice and discrimination are compounded when the dissenter is Negro or one seeking social action on behalf of Negroes. The tendency to oppose, suppress, discredit, exclude, and keep dissenters weak and in their place often parallels the treatment accorded to children by parents, to lower classes by upper classes, to women by men, to students by university administrators, and to labor by management at critical points in their interactions.

As I have stated[5] these roles are much influenced by unconscious primary process thinking:

> "As soon as small children become capable of learning voluntary control of their behavior, they are taught that certain body products like feces and urine, and certain feelings and thoughts, especially sexual or hostile ones, must be excluded from social interaction and kept in their place. Images of these body products, feelings, and thoughts, which have been assigned low value and segregated out of the main stream of social interaction, become associated with each other and with images of any other objects which are also segregated. Members of any segregated group are considered to be dirty, smelly, and undisciplined in aggressive and sexual behavior. The patterns of life imposed upon colored persons in America along with their brown coloring has reinforced the tendency to link in our unconscious minds images of colored persons with images of body products and body processes which must be excluded from social circumstances.
>
> Thus, as long as Negroes are in any way excluded from social interaction they will be linked in the unconscious thinking processes with those ideas, feelings, and body products which tend to be excluded from social interaction as unacceptable. Under such circumstances contact with Negroes will be accompanied by a conscious sense of something forbidden, unpleasant, bad, and possibly dirty or dangerous. Unconscious sexual, aggressive, and other repressed feelings may tend to emerge into consciousness in the presence of Negroes. These emerging repressed thoughts and feelings concerning body products may give the feeling of an influx of alien elements which threaten to flow out of control and make a mess. It may seem imperative that separation and suppression of Negroes must recur in order to maintain order. This explains the intensity of reaction to desegregation or to movement of Negroes into a white community at times."

The Roots of Group Dissent in the History of the Issue

The hospital and ward studies described in Part IV, Examples 1 and 2 indicate that knowledge of the context and historical background of dissent may be indispensable. Nowhere is this more clear than in the dissent of racial activists.

Under slavery Negroes were bonded to whites in a forced, one-sided marriage with great disparity between the roles of the partners. Some forces have pressed to restructure the marriage toward more equality, some have pressed to continue the bad marriage as it is, and some have pressed toward separation or divorce. The essence of the psychodynamics of racial activists lies in an interaction psychology which embraces both white and black thinking. Again:[5]

> Ordinarily, the role of a man has been conceptualized with leadership, wisdom, bravery, power, work, and sexual components. In the white-black relationship, whites were conceptualized with leadership, wisdom, and bravery components and blacks with power, work, and sexual components.
>
> Usually, a woman has been conceptualized as one who is pure, loved, respected, and who has feeding, cleaning, sexual, and other service functions. In the white-black relationship, whites were conceptualized as pure, loved, and respected while blacks were viewed in terms of feeding, cleaning, menial, and sexual functions.
>
> Children ordinarily have been conceptualized as ones who are loved, waited on, cleaned, fed, and also dirty, smelly, untrained, undisciplined, pleasure-oriented dependents who need to be led. In the white-black relationship, whites were related to as the former while blacks were related to as the latter.
>
> Clearly, such components of the role of man, woman and child as supported the master role were ascribed to whites, while such components of the role of man, woman, and child as supported the servant role were imputed to blacks.
>
> In all these roles there is a manifest theme for the white person which states to the black, "you are a child, I am a parent, I will direct you, I will take care of you." The latent theme for the white person never expressed and seldom realized is: "I am a child, you are a parent, feed me, clean up, do the housework and labor for me." The manifest theme of the black person expressed toward the white person is as follows: "I do what I am told, I am a child, you are a parent." The latent theme is: "I do everything you need done, I am a parent, you are a child."
>
> The roles as described and the myths associated with them have seemed appropriate and realistic to those who have not known, or have forgotten, that these roles were designed and forged into personalities under extreme coercion. A body of law, and of common practice, codes, social and psychological pressures upon grown-ups, and child rearing practices with children, have ensured that whites and blacks regularly enact their assigned roles in the American segregation drama.

Racial activists take action to effect social changes in the white-Negro relationship. No clear understanding of the general psychology of whites and blacks can take place without parallel examination of the psychology of both. This may be compared to a conjoint therapy approach. For generations, management of internal impulses, self sacrifice, and self suppression have been emphasized, while care for self, and mastery of external reality were subordinated in Negroes who were relating to whites. For whites interacting with Negroes, care for self, self expression, and mastery of external reality were emphasized, while management of internal impulses and sharing were subordinated.

As the Civil Rights Movement has developed through a succession of stages, civil rights workers as well as others experienced reinforcement or suppression of different patterns in their behavior. Conscience and morality motivated the early "unity," "black and white together," pro-integration phase which mobilized peaceful collaborative Negroes and whites. A phase of "white backlash," with group formation among whites based upon distrust, hatred, and fear of blacks produced an unlabelled "White Power Movement" which prevented significant achievement of original civil rights goals. When "White Power" defeated all the conscience and morality which could be mustered,[6] the Civil Rights Movement became motivated by a counter movement toward power and group formation based on distrust, hostility, separatism, and self-development themes. Each stage appealed to different persons and reinforced different patterns of conduct. Once some group formation, positive identity, improved self esteem, and internal sense of authority developed, white persons were no longer permitted to use unwarranted aggression against a "black brother" in a ghetto situation. Wherever white aggression against blacks occurs in ghettos there is increased potential for black group reaction. In many instances this has developed into riot proportion.[7] The increase in political power and the capacity for well-co-ordinated action which have developed from this group formation process among Negroes will move the interaction into subsequent stages. A focus upon the civil rights *interaction* rather than the Civil Rights Movement will lead to a clearer perception of the issues.

To understand the psychology of the racial activist at work on a social issue one must understand the psychology of the blacks and whites with whom he works. To understand the psychology of the racial activist at work on the psychoanalysts' couch, the "black" part of his personality and the "white" part of his personality must be carefully analyzed. No racial activist and no other dissenter can work on a social issue which does not have its affect-laden intrapsychic counterpart.

Strong dissenters relate themselves to their target issues with intensity and persistence. The psychology of the dissenter is so intertwined with that of those dissented against that both must be examined for understanding. The defensive, suppressive, or retaliatory responses, or suspicions and prejudicial attitudes accorded to dissenters are evidence of general psychological responses to dissenters on part of those dissented against. Often those dissented against respond with counterattack, with invocation of rules and principles which minimize understanding of the dissenter and invite perception of the dissenter as irrational, morally wrong, or unlawful in his behavior. Such a reaction is as irrational as that of parents who reject children or of children who reject parents when there are opportunities for a more adaptive relationship. In the interaction between the dissenter and dissented against, there is need for processes which undermine the resistances to understanding in each and which maximize the opportunity for perception of common goals to provide the basis for such identification with one another as they need for mature interaction. With such processes we anticipate adaptive changes in both mother and child, husband and wife, adult and adolescent, dissenter and dissented against when they are in conflict. All we know about conflict management may be invoked in the management of dissent. When two interacting parties differing greatly in power and resources are in conflict manifested by dissent of the lesser party, some parallel or conjoint process of conflict–resolving activity is essential for a constructive outcome.

Those who ask about the psychodynamics of dissenters may well be asking unconsciously a question which is laden with prejudice. There may be as much or more that is maladaptive or destructive in the party dissented against, yet we do not even have a label for that party. As is so often true when answers are sought via psychoanalysis, the answers are less to be found in the content than in the process. It is quite possible that psychoanalysts are more dissented against than any other individuals as they receive the complaints directed against mothers, fathers, siblings, and organizations across the land. They analyze not only the transference aspects of the dissent directed toward them but the countertransference as well. In the final analysis psychoanalysts may offer more to the understanding of dissent by reference to the processes they use than to the content they have produced. Reference to the processes psychoanalysts use, makes it clear that the party dissented against should receive attention including a label. The thinking and behavior in the party dissented against should be subjected to parallel scrutiny. Only by formulating psychodynamics based on parallel scrutiny of the dissenter and the dissented against can we achieve the order of understanding which we have been gaining about the interaction of mother and child, of husband and wife, of adult and adolescent and other paired complementary interactions wherever they occur.

At times the relatively mature, reality oriented dissenting actions of patients invite attention to social circumstances which seem pathogenic in that they give rise to intense intrapsychic and interpersonal conflict, reinforce exploitative or destructive pregenital relationships, or precipitate symptoms in large numbers of people. In an age of dissent every agent which ministers to people is pressed to examine itself more carefully. Additional attention to countertransference and reexamination of attitude, beliefs, and concepts has prompted some psychoanalysts to reduce their isolation from social issues. By employing modifications of their tools for investigating and understanding behavior, they function as participating observers in social conflicts somewhat as they do in the intrapsychic and interpersonal conflicts of their patients. They may provide valuable assistance by helping to clarify misunderstandings, by identifying inadequacies in communication, by exposing unseen contributions to problems and unnoticed resistance to change.

The analyst as a dissenter on social issues comprises a separate unexplored chapter which should be mentioned although it has not been studied. Many of my patients were aware that I took a stand as a dissenter on a number of issues related to education and to certain social circumstances. Patients occasionally saw press reports or TV programs in which I appeared or in which reference to me was made. In only one instance did a patient show dissenting activity as a manifestation of positive transference. I attempted to apply psychoanalytic concepts and to maintain my professional role in any dissenting activity in which I engaged. I viewed this as a clear responsibility to my patients, since it offered an extension of my relationship with them which might be involved in the treatment. Adherence to objectivity, dispassionate comprehensive appraisal of issues, balanced discussion of dynamics, and interpretations of resistance to understanding or to change at times when helpful effects were possible, was my conscious goal at all times. I had no direct contact with psychoanalytic patients outside of the analyses, but did serve on committees with two patients concurrently in treatment with me, and three of my ex-patients (including one ex-analysand) with no apparent adverse effects.

It may be useful to remind ourselves that the psychodynamics of dissent could be examined by studying psychoanalysts instead of patients. Sigmund Freud expressed dissent via his contributions and set in motion a chain reaction of psychodynamics which includes all analysts. As his contributions gained support and influence, the development of dissent against Freud by individuals and later by groups has been of critical importance to the development of psychoanalysis as a science.

REFERENCES

1. Hoffer, Eric: The True Believer. New York, Harper and Row, 1951, p. 21.
2. Marmor, J., Bernard, V., and Ottenberg, P.: Psychodynamics of Group Opposition to Health Programs. Am. J. of Orthopsych., Vol. 30, No. 2, April 1960.
3. Erikson, E.: Childhood and Society. New York, Norton, 1963, pp. 247-274.
4. Bowlby, J. B.: The Nature of the Child's Tie to His Mother. Int. J. Psychoanalysis, 39: pp. 350-373, 1958.
5. Pinderhughes, C. A.: Pathogenic Social Structure: A Prime Target for Preventive Psychiatric Intervention. J. Natl. Med. Ass., Vol. 8, No. 6, pp. 424-429, November, 1966.
6. Mendelsohn, J.: The Martyrs. New York, Harper and Row, 1965.
7. Pinderhughes, C. A. and LeVine, H. O.: The Psychology of Adolescents in a Peaceful Protest and in an Urban Riot. (Paper presented at New York State Divisional APA Meeting, November 17, 1967.)

Discussion by C. Eric Lincoln, Ph.D.

Dr. Pinderhughes has provided new windows through which we may examine a very interesting, and what is to some a very perplexing psychosocial phenomenon. If the 1920's and the post-World War I era constituted the "age of anxiety," certainly, this is the age and the era of dissent. On the surface at least, dissent seems to be extremely widespread, with the dissenters sometimes getting their lines confused and dissenting against each other and even against dissent.

The meaning of dissent. Dr. Pinderhughes describes dissent as "a tool for updating the life style of a culture, of a family, of a group, of a society." "The essence of dissent" he finds to be "opposition expressed from a position which is considered inferior in power, in numbers, in substance." There is good sociological evidence for that power in the control over decisions. Negroes have no power, hence the civil rights dissents. Students have no real power; Vietnam protestors feel they have no power, etc.

However, recent sociological studies also indicate that the typical student dissenter, (at least) comes from the "establishment," where the power is *poten-*

tially his. Again, with the notable exception of black dissent, much of the dissent in this country is carried on by persons who are not themselves subject to the strictures against which their protest is directed. Whites in the Civil Rights Movement, deferred students, clergymen against Vietnam, and women protesters are readily available examples.

The *psychology of the dissenter* is projected by Dr. Pinderhughes as a spectrum of dissent which includes: (a) "the true dissenter" who is motivated primarily by feelings activated by the issue; and (b) the pseudo dissenter, who employs dissent as a means to an end which lies outside of the issue. Presumably, there are graduations of ambivalence in between. These findings seem to be amply illustrated in the Civil Rights Movement, especially in much of the Student Activist Movement.

From another perspective, dissenters may be classified as: (a) the "Transformationists" who are willing to risk everything to change society, and (b) the "Recessionists" who have despaired of the possibility or the desirability of changing society. They retreat in the face of an insuperable power and conclude that the society which has overwhelmed them is not worth the effort required to salvage it.

Who are the dissenters? Dr. Pinderhughes' clinical experience reveals that "more individuals express dissent in periods of reduced stability." On the basis of this and supporting observations he is able to offer two hypotheses:

1. "When a social order is relatively stable and dissent is endemic, the number of patients actually engaged in dissenting social action is apt to be small and the psychodynamics of those patients relatively circumscribed."

2. "When dissent exists in epidemic proportions, the number of patients actively engaged in social action is apt to be large and the psychodynamics of those patients quite varied."

There is a good deal of evidence that dissent may be one way of reducing anxiety. For example, the demands for counseling services are likely to be dramatically reduced on a college campus during a period of vigorous, activistic dissent. When the dissent has run its course, there is usually a very sharp increase in the number of students seeking counseling or other forms of therapy. It is my expectation that this experience will be confirmed as repeated in the larger society.

Secondly, we can further identify those people in Dr. Pinderhughes' hypotheses who are likely to be involved in dissent. On the basis of recent findings at Berkeley, using the Mean Standard Omnibus Personality Inventory Scores and other instruments, we can predict that they will come from the upper-middle class; they will score significantly higher than nondissenters in introversion, complexity, estheticism, autonomy, impulse expression and religious liberalism. But their predisposition to anxiety will be even more dramatically marked (James W. Trent and Judith L. Craise. "Commitment and Conformity in the American College." Journal of Social Issues, July, 1967.).

The racial factor. This introduces other dimensions and categories of behavior which do not apparently conform wholly to the general findings in either the psychology or the sociology of dissent. There seem to be abundant evidence, for example, that the black intellectual, because of his social and professional marginality has always been a prime candidate for schizophrenia or some other psychosis. Yet, the black middle class has not been particularly noted for vigorous dissent. On the other hand, the college-aged sons and daughters of the black middle class are precisely the progenitors, of this era of dissent which has seen their rebellion adopted and broadened by white youth, and by extension, their parents. Much more study needs to be done in this area to determine to what degree, if any, the frequency, nature, intensity and duration of dissent are affected by racial factors.

Dr. Pinderhughes' analogy which sees the child-parent conflict over rights as being the prototype of the civil rights conflict is an interesting one. However, the analogy breaks down in that in the prototype the child inevitably grows up, becomes an adult, and exercises power. There is as yet insufficient evidence to project the analogy to its final and logical stages with reference to what is proposed as its ectype, for in the American drama of race relations, the black American never grows up, becomes an adult, or exercises power. Again, the theory that dissent presupposes an abject powerlessness, while it applies to the black situation without serious qualification, may not accurately have universal application. Certainly there is a difference in the power and the power-potential of the handful of black students in SNCC and the power, real or potential, of the scions of the establishment who are based at Berkeley or Columbia. I am not prepared to offer a resolution of this seeming discrepancy, but there is no doubt in my mind that it exists.

Dr. Pinderhughes suggests that in the unconscious of white Americans the "brown coloring" of black Americans is associated with body products. In the black in-group just the reverse is true. A person with a very light or white skin is often referred to as "fecal-colored," or its equivalent. While this has no necessary relation to racial dissent, it does suggest that Dr. Pinderhughes' theses regarding this kind of projection probably has extensive application among different cultural or ethnic groups, and that it serves to confirm the ideal self image of the subject.

Finally, the interaction psychology which Dr. Pinderhughes sees as illustrating white-black conceptualizations is particularly important at this juncture of racial relations when there is a sharply increasing polarity between racial groups. Whites have been traditionally associated with positive values such as leadership, wisdom, bravery, purity, love, respect, etc.; blacks with lesser values usually having physical or physiological connotations. At the heart of the traditional conceptualization, the white man considered the black man a child—a dirty, ugly, stupid child at that. It would be interesting to see ten years from now what implications there will be for the psychodynamics of dissent in the new mood of the black revolution—where in the black in-group there is an inversion of conceptualized roles and values—and where in the white in-group there is at least the stirrings of doubt about a conceptualization that has always been taken for granted. This challenges further research.

PSYCHOLOGICAL ISSUES IN THE DEVELOPMENT OF YOUNG RADICALS

By KENNETH KENISTON, Ph.D.

VIETNAM SUMMER was a large and far-reaching group dedicated to organizing new constituencies to oppose American involvement in Southeast Asia. Those who conceived of the project, like those who actually led it, defined themselves as radicals, as members of the New Left, or the Movement for Social Change. They were strongly opposed to the war in Vietnam, but many were equally or more interested in other major changes in American policies, both domestic and international.

This chapter is focused specifically upon a dozen young men and women who had come up through the Movement: New Leftists with a deep commitment to community organizing and peace work as a part of a broader program of social change. In their early or mid-twenties, they were committed radicals, most of whom had already devoted several years to full-time work in the Movement.

My first interview focused largely on their political activities and ideas as these had emerged in adolescence and after. The second interview was an exploration of their early family environment. In third and subsequent interviews I tried to establish connections between early life and tradition on the one hand, and current political views and activities on the other. This report is limited to observations on "committed" radicals who consider themselves part of the Movement, and spent at least one year during which their primary work involved work in community organizing, civil rights, New Left coordinating or peace.

The age of those interviewed ranged from 19 to 27, with an average of 23. All were white; three were women. Two came from lower middle-class families, two from upper-class, the remainder from upper middle-class; most frequently, their fathers were business men, teachers or professionals. Seven came from Protestant backgrounds, five from Jewish

---

Supported by the Foundations Fund for Research in Psychiatry.

backgrounds, and two from Catholic families. Three came from clearly "radical" families, with parents who had been active in left wing movements during the 1930's; the parents of most of the remainder were Democrats, often of a "Stevensonian" persuasion. All of the interviewees had at least begun college, and the colleges they had attended were generally academically excellent, highly selective liberal arts colleges or private universities. Socioeconomically, these young radicals come from relatively privileged and advantaged sectors of American society. They were clearly not "representative" of all New Leftists; nor could I, from the information gathered during the summer, judge in any systematic way how they might differ from other young radicals who might have been interviewed in other contexts. Indeed, I am still not clear how one would go about finding a "representative" group of young radicals.

For each subject I had an average of approximately five hours of interview material, plus, usually, an equal amount of time in which I had observed him at work, in meetings, at national conferences, and so on. While the amount of information at my disposal is extremely limited as compared to that which could be garnered from an intensive psychotherapeutic relationship, the unusual candor and introspectiveness of this group made me feel that, in most instances, I was able to distinguish fact from self-justifying myth and obtain a relatively accurate picture.

By focusing solely on the psychological themes that were found in these radicals, I omit other issues of great importance. In particular, I will not consider here the formative effect upon these young radicals of their average of two or three years of experience in New Left organizing work, nor will I be able to touch upon the equally important topic of the impact upon them of social and historical forces that contributed to shaping their lives. It should be understood, then, that in discussing psychological issues in radical development, I am arbitrarily separating out one set of determinants of radicalism, and deliberately neglecting other group, experiential and historical factors that have equal importance.

Two Inadequate Hypotheses

In discussions of the New Left, two hypotheses are continually put forward; I will call these the "radical rebel" hypothesis and the "red diaper baby" hypothesis.

The radical rebel hypothesis is clearly most widely held in the general population, and sometimes advanced even by scholarly observers of the New Left. According to this thesis the position of today's young radicals reflects a violent rebellion against and hatred of all male, parental, and societal authority. The radical, according to this view, is "displacing" the conflicts of his family onto society and the world, "acting out" intrapsychic conflicts in his external behavior. A more sophisticated version is the view that radicals tend to come from families in which they have experienced, perhaps to an unusual degree, the ultimate barrenness, flatness, and emptiness of American middle-class life. Specifically, sympathy and even pity towards parents is combined with a major effort to avoid lives seen as empty, spiritually impoverished, and often "quietly desperate."

A second, sharply contrasting hypothesis was often put forward by individuals active in New Left work. This thesis could be called the "red diaper baby" hypothesis, a phrase used within the Vietnam Summer office to characterize those workers who came from families with a left wing tradition. According to this hypothesis, present day radicals come largely from politically radical families in which, from early childhood, they have been exposed to radical ideas about social reform and political action. The development of the radical as a person, then, is portrayed as smooth, continuous and uninterrupted, as a simple assimilation of parental values of dissent and indignation at modern American society, coupled with a determination to work toward correcting these injustices.

In support of this hypothesis Bettina Aptheker, the radical daughter of well-known Communist parents, and several highly publicized figures are frequently cited. However, neither the radical rebel nor the red diaper baby hypothesis proved relevant to the lives of those I interviewed. Both theories of radicalism overlook the actual complexity of human development, and each posits the possibility of either a total break with the past or a total acceptance of it which rarely if ever occurs in life.

The Schism in the Paternal Image

The patterns of relationship with parents found among the young radicals I interviewed were highly diverse. On the one hand there were a few, particularly those from left wing backgrounds, who described idyllic relationships with their parents: close understanding, support, and mutual sympathy were the rule throughout childhood and continued with only occasional interruption during adolescence into the present. At the other extreme were a smaller number whose development had been marked by violent rebellion against parents, rebellion which they invariably connect explicitly to their current involvement in radical politics. But perhaps the most striking general feature was great understanding of the parents, with relative detachment, compassion and ability to view the parents "in the round," as themselves complex, differentiated human beings. And perhaps related to this current sympathy was a portrait of early family relationships and a family atmosphere most often characterized by considerable parental warmth, sympathy, understanding and idealism. Put into an over-simplified formula, these young radicals more often came from "happy homes" than not.

In addition, a number of more specific themes recurred. A great variety of anecdotes, early recollections and later events suggest an unusually strong tie between these young men and their mothers, in the first years of life. Such a close maternal tie is of course not unusual amongst young Americans, but in many of these individuals it seems to have been particularly intense, and to have evolved into an unusual sensitivity and responsiveness to the mother's wishes.

Another common characteristic of the mothers was their eagerness for intellectual achievement for their sons. The pattern is well described by one interviewee who mentioned his mother's semijoking attitude when he returned home with a grade of 98: "Where are the other two points?" Most often, the fathers seconded the maternal emphasis on intellectual attainment, and the sons complied by doing extremely well in academic work. Whatever the specific way in which maternal interest in intellectual achievement was transmitted, one of its corrolaries in later life seems to have been the frequent identification of intellectuality with conformity to maternal pressures.

More central to the overall development of these radical activists, however, is their intense and highly ambivalent involvement with their fathers. The term "ambivalent" somewhat understates the issue, for ambivalence toward fathers and mothers is routine, expectable, and unremarkable in the development of youth. With these young New Leftists, however, ambivalence towards their fathers was greater than that found in most young men and women: the image of the father, as seen through interviews, was radically split into two contradictory parts. On the one hand, the father is frequently portrayed as highly ethical, intellectually strong, principled, honest, and politically idealistic. But on the other hand, the same father in other contexts is sometimes seen as a passive, humiliated, unsuccessful, weak, acquiescent or otherwise inadequate person. This schism in the paternal image was found in almost all of those interviewed and seems of great importance in their development. Examples may help clarify the complexities of relationships to the father.

> One subject remembered very positively his long walks with his father during early childhood: the two would discuss social and political issues, and in this context, he saw his father as effective, competent and highly idealistic. But he viewed his father as being dominated and tyrannized over by his mother, as unnecessarily acquiescent and passive, and as strikingly ineffectual in his work. Another interviewee discussed at length his early hatred of his father, and his later violent and successful rebellion against him. But when asked about the sources of his political beliefs, he cited first his father's "honesty" in recognizing the "fundamental illegitimacy" of the business system in which the father was extremely successful, and went on to suggest his strong identification with his father and his father's youthful questioning and rebellious attitude. Still another interviewee considered his father very much of a "failure," given his father's considerable basic talents, and had tried very hard not to follow in his father's footsteps; at the same time, he was much identified with his father's values of enterprise and constructive work

Family Orientation

Within this context of an intensely ambivalent relationship to the father, there was a continuum of family relationships, and especially paternal relationships, along which most activists could be placed. At one end are those few from relatively "radical" political backgrounds who convincingly describe a family milieu of unusual warmth, demonstrativeness, openness and congeniality; at the other end is an equally small number of New Leftists who describe early family environments of violent conflict or extraordinary isolation between family members.

In both types of families, psychological development had special characteristics. For example, some young radicals described families which seemed extraordinary in their deveotion to the independence of their children and their dedication to the highest moral principles, but their own developments suggested that such families may be in a subtle way particularly difficult for a capable son to cope with. Despite the strength they can give a growing child, they provide him as an adolescent with little to rebel against, and the attainment of independence and adulthood may become paradoxically synonomous with passive compliance to family wishes.

Conversely, families that gave their children a great deal to rebel against always seem to have provided something to emulate as well.

For example, one son engaged from early life in furious battles with his father which clearly won his father's grudging respect; thereafter there existed an unstated but strong bond between father and son, perhaps because each saw in the other a fighter like himself. In other cases, the father's domestic ineffectuality was partly offset by his intellectual effectiveness, his high principles or his more effective past. In still another instance the son repudiated his father's humiliating failure, but identified strongly with his father's and grandfather's values. For such young radicals, overt repudiation is frequently coupled with identification, and by the time they were interviewed, most were able to acknowledge not only the negative but the positive lessons they had learned from their fathers.

These complex father-son relationships suggest that neither the "red diaper baby" nor the "radical rebel" hypothesis is adequate, although neither is totally incorrect. Like most stereotypes, they indicate one small part of the truth, while omitting the crux of the matter. The vast majority of those interviewed were in some respects *both* red diaper babies and radical rebels: their development involved a simultaneous identification with one aspect of the father, particularly with the father's principles, goals and ideals, and an equally frequent rejection of another aspect of the father, particularly the father's ineffectuality, inactivity, subservience or apparent failure.

Given the very real differences in the family situations of each of those interviewed, it is obvious that no one characterization will suffice to describe the relationship of the radical's parents to each other. Again, there is a full range from open parental hostility, occasionally verging on divorce, to a more common pattern of intense, close, mutuality and sharing between the parents. Yet above and beyond these differences, virtually all of these families shared a common focus on morality, defined not as the narrow restraint of impulses, but as a higher ethical orientation toward personal relationships within and without the family.

Almost without exception, then, these young radicals came from families with *principled parents*. They were almost invariably disciplined as children by "psychological methods". Yet this term does not do justice to the role of principle in their families. Somehow their parents communicated, often without ever saying it outright, that human behavior was to be judged primarily in terms of its conformity with general ethical principles, that right conduct was to be deduced from general maxims concerning human relationships, and that what mattered most in the evaluation of individual worth was the ability to act in conformity with such principles. In some of these families, the principles were specifically religious, as with more than half of the interviewees who grew up in families with explicit religious affiliations. For a smaller number, the principles were primarily ethical and cultural, particularly amongst young radicals from secular Jewish families with a strong ethical identification with Judaism. And in a few instances, the principles were primarily philosophical and political, rather than religious or cultural. But for virtually all of these young radicals, the atmosphere created within their early families was one in which ethical principles occupied the highest positions, and in which human events and human conduct were automatically judged in accordance with these principles. This familial orientation to principle is an important continuity in the lives of these young radicals.

The Early Sense of Specialness

The pre-adolescent experience of these young radicals again varies enormously, depending on family, social and historical circumstance. What distinguishes virtually all of them is a feeling, from an early age, of being somehow different, separate, apart and exceptional, a self characterization which they viewed ambivalently and which continues to be important throughout their later development. The sources of this sense of specialness are varied, but cluster around a few crucial themes. For one, these young radicals frequently come from families whose style, ethnicity, or ideology separated them from the surrounding community, a separation which contributed to an early sense of "alienness" from others in their environment. The sense of specialness was often intensified further by early intellectual excellence, a capacity which usually placed them in the scholarly "elites" of their grade and secondary schools. Moreover, many of the interviewees were leaders from an early age, especially in the defense of unpopular positions. One, for example, recalls leading a disastrously unsuccessful charge of the bullied children in his school against the bullies. Despite the catastrophic failure of this early revolutionary effort (the bullies triumphed as usual), he now consciously relates this act to his continuing interest in organizing the oppressed. Another interviewee recalls being beaten up in grade school because of his parents' political views; still another, whose family represented a minority group in their neighborhood, was ostracized and bullied because of his minority group status.

Another frequent source of an early sense of uniqueness was the apparently precocious political, social, and historical consciousness of these radicals as children. For several, the first memory is that of the end of World War II; another's earliest recollection involves a political figure who was mobbed by a hostile crowd; still another recalls his early bewilderment at his own favored social position in the community.

A more explicit political and social consciousness seems generally to have developed during the early grades of school. Many report holding "unpopular" views and defending them vigorously, usually with the support of their parents. In some instances, these students were the "leaders of the opposition" around such issues as grade school drills for atomic attack, local elections, or presidential campaigns. Even when asked, they can rarely recall favoring a "majority" position: e.g. they supported a Democratic candidate in a strongly Republican area, advocated peace in a bellicose community and defended the underdog.

My emphasis on a sense of specialness, on early intellectual ability and moral precocity, should not be taken to suggest that these were unusually studious, withdrawn, or joyless children. The opposite seems to have been the case: many were good sandlot athletes, they had good friends and playmates, and they were often elected president of their grade school classrooms. With few exceptions, they describe themselves as sociable and outgoing, as "making friends easily," and as moving readily in the social world of school and family. The overall picture, then, is one of a relatively happy and successful childhood, of general effectiveness in school, and of a globally pleasant recollection of the preadolescent years.

Turmoil-Filled Adolescence

At about the age of 13, a drastic reversal in outer behavior, paralleled by intrapsychic changes and conflicts, frequently occurred. The preadolescent pattern of outgoing activity was transformed, often in a few months, to a new style of seclusiveness, a feeling of social awkwardness and a sense of moral inferiority, coupled with intense intellectual concerns and at times extreme religiosity. The outgoing happy child was transformed, as it were overnight, into the shy, awkward, and tormented early adolescent.

Although such changes in behavior and inner life during adolescence are described as typical by European commentators on adolescence, they are in fact relatively infrequent in the development of young Americans. The more usual pattern for American youth involves turning away from the family toward the peer group, immersion in its activities, and the use of one's agemates as a buffer against one's own impulses and ties to one's family. In a sense, then, many of my subjects experienced a more "European" adolescence than their typical American contemporaries: certainly the ascetism, guilt, and defensive intellectualization of these young men fits the classical description of early adolescence of upper middle-class European youth.

The focal and generally quite conscious issue during this period was the emergence of sexuality. Again and again, those interviewed explicitly related their early adolescent conflicts to feelings about sex, about their own erotic fantasies, and about the changes in their bodies. How much of this was fully conscious at the time we cannot be sure; but now, the relationship is presented as direct and unequivocal. The actual symptoms of inner turmoil varied considerably from individual to individual.

> One interviewee elaborated a private fantasy kingdom, complete with aristocracies, masses, and revolutions, which at times became more important to him than his other activities. Others reported extraordinary self consciousness and shyness, especially toward girls with whom they had once felt easy, relaxed and carefree. Other interviewees became intensely and scrupulously religious in early adolescence, highly involved in church or temple activities, and in one case, elaborating a private set of obsessional religious rituals. A strong propensity toward ascetism, toward rigorous self discipline and toward an effort to "deny the flesh" was related by several to guilt over masturbation and sexual fantasies. Most of those interviewed reported that at this time of their lives, they had become intensely concerned with abstract philosophical and moral questions, and had spent much of their time in voracious readings, journal keeping and theorizing.
>
> In early adolescence, too, feelings of loneliness, solitude, and isolation came markedly to the fore. Several interviewees began adolescent diaries which they kept for many years, prefacing them with thoughts such as, "Since I have no one to talk to, I will have to talk to myself." Turning towards themselves rather than towards the peer group, these young radicals began a habit of self-analysis whose results were evident in the insight with which, years later, they could discuss their lives. Yet a journal is rarely an adequate

companion, and feelings of painful loneliness were common. Sometimes because of an explicit feeling that they were especially filled with evil thoughts, and often because they felt "different" from their contemporaries, they were forced back on their own resources. In time, necessity became a virtue for some: they learned to occupy themselves, and not only to tolerate loneliness but to need solitude.

Some time after this drastic reversal of behavior, and the development of ascetism and intellectualization, many of the interviewees also entered into a period of violent rebellion against their parents, usually focussed around parental "unfairness" and "injustice". The particular issues at stake in these mid-adolescent rebellions were not in themselves unusual: they center largely around the individual's view that his parents attempted to restrict him excessively, not allowing him sufficient freedom, attempting to control his life, to plan his future, and so on. But the intensity of the rebellion does seem in many instances unusual, as does the feeling of moral outrage and betrayal that accompanied it. Again, the issue of principle is important: in at least some of these families, parental preaching was not practiced when the chips were down; in particular, these adolescents felt outrageously betrayed when the same parents who had consistently encouraged them to be independent, freethinking and autonomous now intervened in adolescence to attempt to direct their lives.

In a few subjects the turmoil of early adolescence eventuated in the development of psychological symptoms during late adolescence which, had they come to the attention of a psychiatrist at the time, would have been considered ominous. They included, for example, the previously mentioned elaboration of a fantasy world more real and engrossing than the actual world, the emergence of elaborately obsessional fears of sexual intimacy, an abortive suicidal gesture, and the development of a transient paranoid view of the world. Most of those interviewed, however, developed no overt symptoms, but describe themselves as often unhappy, subjectively quite isolated, and inwardly frustrated and unfulfilled despite outer success.

Yet to discuss merely the turmoil, the frequent unhappiness, and in a few cases, the psychological symptoms of adolescence would be to overlook the continuing psychological strength shown by these young radicals. Whatever their inner turmoil, whatever their private misery, and despite symptoms that would have incapacitated many, these young men and women "functioned" extremely well. Indeed, those with the most intense conflicts or the most apparently crippling symptoms often seem to have been the most outwardly successful. The early adolescent period of seclusiveness and withdrawal was gradually overcome; relationships with the opposite sex were generally established, painfully and slowly but surely; and the pre-adolescent pattern of leadership coupled with intellectual achievement was generally resumed. In their college careers some of those interviewed described themselves initially as "typical college students," but further inquiry usually revealed that they had been outstanding scholastically, editors of college publications, or had held positions of leadership in campus organizations with a social or political orientation.

The End of the Line

As they initially talked about their entry into Movement activities, those interviewed described a singularly gradual, undramatic, and uneventful development. There were no conversions in this group. Instead, interviewees spoke of a slow process of increasing political involvement, each step of which seemed, inevitably, to lead imperceptibly to an emerging sense of self as a radical. These young activists almost uniformly found it impossible to identify any one period or issue around which they came to think of themselves as radicals. Nor could they select any particular experience as decisive to their political careers.

Yet seen in the context of their overall psychological development, the gradual movement toward increasing involvement in the New Left seemed related to a growing discontent and inner frustration despite outward success or the possibility of it and to an often conscious search for alternatives to their current activities and life styles—in short, to a subtle feeling that they "were nearing the end of the line." This sense of increasing frustration did not stem from failure or rejection: what one interviewee called "the Establishment options" were clearly open to all of these young radicals. They were successful, well organized students, often leaders of their contemporaries, always intelligent and capable; their superficial problems was that of a surfeit of options. But subjectively, none of these options seemed particularly interesting to them, nor in many cases were they experienced as psychologically available at all.

Most often during the college years but sometimes before and sometimes after, these young radicals seem to have reached a point where the "Establishment options" became psychologically exhausted; and it was generally at about the same time that they began to turn, gradually and unreflectively, toward increasing commitment to the Movement for social change. One interviewee, headed toward the career which his parents desired him to enter, began to do extremely poorly in his required courses, and out of the ensuing crisis, moved gradually into radical activities. Another, after a valiant effort at more conventional political involvement, felt an increasing sense of self-defeating frustration, became seriously ill for several weeks (a fact to which he attributes psychological meaning), and then turned toward involvement in more radical political activities.

In the development of a sense of having reached the end of the line, one particularly central issue is specialness, guilt and self doubt. Almost all retained a feeling of possessing special talents and abilities coupled with an unusually high sense of personal principle. For these young men and women, more than for most of their contemporaries, it was important to be doing things that were consistent with principle, to be people who not merely "had principles," but whose lives embodied and implemented their principles. The increasing sense of the inadequacy of the Establishment options was connected to a growing conviction that these options were ethically inadequate, as was the life trajectory that headed one toward them. In discussing their growing feelings of self dissatisfaction, their feelings that they were "wasting life," that what they had been doing was in some way "self-destructive," the issue of principle and the profound guilt and shame that arose from failing to follow its dictates is crucial.

In the gradual evolution of a commitment to the Movement, then, many factors were fused, with varying weights depending on the individual and his circumstances. Perhaps the childhood sense of specialness made it more psychologically possible for these young men and women to turn toward a Movement which represents a special minority of young Americans. Certainly the role of principle, and specifically the frustrations and guilts of an unprincipled life, played a crucial role for all. But beyond this, the sense of reaching the end of the line also reflected different psychological issues for each person: what they shared was merely the growing explicit awareness that their present lives were inadequate.

The Growing Sense of Rightness

The brief years that passed since the first development of a radical commitment brought many different fates to those interviewed: organizing work in the slums, interruptions and recontinuations of college careers and graduate studies, civil rights work in the south, teaching the children of migrant workers, developing groups of southern white students who support civil rights, disappointments, defeats, and occasionally a rare success. But whatever their activities in the ensuing years, the central psychological theme of these years was a growing sense of rightness, in both meanings of this ambiguous term. On the one hand, these young radicals increasingly felt that what they were doing was psychologically "right" for them; in accordance with their needs, responsive to their talents, and consistent with their innermost fantasies. At the same time, "rightness," not to be confused with fanaticism or self righteousness meant, a conviction that what they were doing, though frustrating, difficult and filled with hardship and triviality, was for them unquestionably in accord with their own highest principles.

This confluence of psychological and ethical themes helped give to the activities and apparent "choices" of these young radicals a peculiar quality of automaticity and naturalness. For example, one individual reported seeing a group of civil rights picketers and joining them. He did not stop to think why, but recalls only that this act seemed natural and inevitable in view of his religious upbringing which had taught him to "bear witness" to his beliefs. When I noted that most of his co-religionists would not have borne witness in this way, he agreed, but insisted that at the time it had "just seemed a natural thing to do".

Again and again, young radicals used similar phrases to characterize what from the outside would seem a decision or a choice: psychologically, the perceived rightness of Movement work virtually removed the need to choose. Such a sense of not choosing but acting "naturally" bespeaks a powerful fusion of conscious and unconscious motives in the service of the developing identity of a radical; specifically, it suggests that for these highly principled young men and women, a new harmony was being established between will and conscience, between ego and super ego; between self and principle.

Growing commitment to radical activities rarely involved any major change in the individual's values and principles. All had been brought up to believe that human beings are equal, that prejudice, hatred, and discrimination are wrong, that suffering should be alleviated, that peace, justice, and equality are admirable values, and that violence should be minimized; therefore the values of the New Left were not at all alien, and becoming a radical required only resolving to act on them.

In this light, the argument that certain kinds of experience such as civil rights work in the South have a "radicalizing effect" on the *values* of the participants is basically incorrect. Insofar as my subjects are representative, their values and principles remained unchanged. What did change, however, was the perception of American reality. Movement work helped bring home the extent to which their own values, by and large the values of our society, were simply ignored in large parts of this country, as in many of our dealings with other nations. What also changed was the awareness of how difficult it is to persuade others to implement the values they profess. And finally, Movement work clearly led to an increased understanding of the tactics and strategies which might be effective in producing the social changes necessary to harmonize American practices with American principles.

As involvement in radical activism increased, the "hang-ups" of earlier adolescence disappeared or became minimal. Obsessional symptoms vanished, leaving in their wake only an unusual capacity for single-minded dedication to a task. The leaden depressions of the past became the realistic discouragements of the present. The fantasy world of knights and peasants was transmuted into the actual world of the New Left in America. Those who came from a radical tradition became increasingly able to accept its full burden without being overwhelmed by it; those in manifestly violent rebellion against their fathers became more able to perceive as well the positive components in their identifications with them. So, too, the negative pole of the sense of specialness, the fear of being especially wrong, especially guilty, or especially unworthy, was transvalued into a more positive sense of working in a special Movement.

In the continuing development of these young radicals, what is impressive is the integrative, "synthetic," adaptive, and personally satisfying nature of the evolving identity. They were able to establish continuity with some features of their parental tradition, while repudiating without excessive conflict other aspects. These were young men and women with a capacity for commitment, for work, for love, and for play. They learned to tolerate frustration, arduous work, and even defeat. Indeed, in the short run at least, "winning" is no longer necessary for them: what is important is the continuing effort to create new tactics, institutional forms, and intellectual formulations that can help transform the modern world. The "identity" of a radical has so far been a highly integrative, satisfying and successful form of personal development.

Yet merely to emphasize the high degree of adaptation and psychological synthesis in these young radicals is not enough. The conflicts of the past are still often visible in the inner and outer tensions of their present lives. There is invariably some unresolved conflict—the reaction against the self abasement of a father, a heightened identification with the ambitions of a mother, or the only-partially successful effort to overcome the sense of superiority which is one side of specialness. To call this a "conflict-free" group would be an overstatement; it would be more exact to say that they have been unusually able to find or create lives and a sense of themselves which enables them to live out their inner strengths, minimize their inner weakness, and reconcile their ambivalences. Yet there is much in these young radicals that does not conform to our view of "typical" development after adolescence. While they have unusually strong commitments, they have few specific plans; their statements about their own futures are invariably vague and hedged about with conditions: "If this were to happen, I would do that. Unless such-and-such occurs, I will probably do this." Despite their emerging sense of self and role, identity formation in these young men and women is far less complete than for most of their contemporaries. They expose themselves to new experiences in deliberate efforts to create the conditions for personal change; they have not "settled down" nor do they plan to. When asked how they visualize themselves 20 years from now, they invariably respond with, "It all depends . . .", and turn to a discussion of the social and political future of America and the world. One might view these young radicals as suffering from a "protracted adolescence," still immersed during their mid-twenties in an "unresolved identity crisis."

Yet before such judgments with their pejorative connotations are sealed, we should recall the special involvement of these young men and women with social and historical changes which have yet to occur. Leaders in a Movement without leaders, "revolutionaries" who believe their revolution will take a generation or more to achieve, they have linked their emerging identities to an emerging movement for social change which may never occur. Given this fusion of inner identity with ongoing historical process, it follows that their development *must* be more flexible than that of their peers who have joined their selfhoods to more conventional and circumscribed tasks. Thus it may be that the radical always "stretches" his development over a longer period than his adjusted contemporaries, and that his identity can be achieved, if ever, only with the sociopolitical transformations he seeks.

Discussion by Robert J. Weiss, M.D.

I agree with Professor Keniston that there are no special psychological characteristics of development or of personality which identify the young radical from the rest of society; I also, with him, reject the notion that the activity of the young radicals is merely an acting out process of unresolved unconscious conflicts from earlier life.

To add confirmatory data, I examined the MMPI scores of some 34 students on campus who were identified by reliable sources as political activists of radicals. A few years ago, I used the device of the number of scale scores on the MMPI two standard deviations above the mean to categorize students who were considered to be not emotionally impaired.[1] If a cutoff point three or more MMPI scale scores 70 or over was used, 16.1 per cent of the entire class of 1967 could be considered emotionally impaired. Applying the same criteria, a review of the MMPI scores of the 34 students seen as activists on the campus currently revealed that 7 students (20.5 per cent) have 3 or more scales scores 70 or over, to show emotional impairment. The difference of 4.4 per cent is not significant since the cells are extremely small. It is of some importance to note that the MMPI profiles revealed no significant pattern difference from that of the class as a whole for those students classified as activists. Six of the students were classified as radicals and of these, 2 had 3 or more scale scores 70 or over and none had highly disturbed MMPI profiles. The distribution of the group by the simple cutoff of 3 or more scale scores 70 or over reveals a distribution remarkably similar to that of the total classes studied in 1967 and for their own classes as well.

The Educational Research Office at Dartmouth computes a predictive index and ranks by percentile all the students entering a class admitted to Dartmouth. This predictive index is based on verbal and mathematical college entrance examination board scores, CEEB achievement scores, secondary school record, rank in class, and a weighting factor for the size and rigorousness of the secondary school. An examination of the percentile rank in class by the predictive index of the 34 students indicates an almost uniform distribution from the top to bottom of the class with 5 students ranking in the 1st to 10th percentile and 5 students in the 91st to 99th percentile. Here again test scores on achievement and standard measures of intellectual capacity reveal no differences in this group of students from that of their classmates. In brief then, a review of other criteria of personality differences or traits of the students involved in activist activities in one school confirms in every detail Professor Keniston's observations of the lack of specificity in personality traits in this group of students.

The major characteristic common to all of them seems to be a concern with human values which is different from those of other students and which certainly by observation of our students seems to be positively correlated with the kind of family background regarding the concern for human values held by their parents. Our students also had liberal parents with a preponderance clustered in the helping professions such as medicine, social work, and the law. Such people have always been part of society, and there is no indication that there is any current excess of students with social concerns. The further issue is the emergence of these students as professional activists after college. Since no explanation for this emerging role can be found in personality traits, conflicts, or backgrounds, it would seem that other areas must be explored. There have been other times in history when similar groups have emerged. One example is the abolitionist movement of the 1830's, composed of young people with special concerns and extreme commitment to a cause. David Donald in a book entitled Lincoln Reconsidered has a chapter on the abolitionists in which his description of the backgrounds of these young people is remarkably similar to that in Professor Keniston's and our own studies.[2] Another instance is the development of professional young people from the middle class concerned with activism in the organization of labor unions, or the dedicated anti-facist group which also provided the well-known Lincoln Brigade in the Spanish Civil War.

Activism focuses around human concerns and human values, but the role played by young activists differs at different times in history. "Why now" is indicated by Professor Keniston in his introductory explanation as to how a group of students asked a concerned but obviously sympathetic individual to investigate their motivations in depth, possibly to combat the obviously fallacious picture of their activities as psychologically deviant.

References

1. Weiss, R. J., Segal, B. E., and Sokol, R.: Epidemiology of Emotional Disturbance in a Men's College. J. Nerv. Ment. Dis. 141:2:240-250 (August) 1965.
2. Donald, David: Lincoln Reconsidered. New York, Alfred A. Knopf, Inc., 1956.

Part III. Differential and Regional Studies

THREE REBELLIONS: RED, BLACK, AND GREEN

By LOUIS JOLYON WEST, M.D.

AND

JAMES R. ALLEN, M.D.

REBELLION IS ASTIR among the younger generation in America today, just as it has been in every generation since and before the Golden Age of Athens. As always, the rebels represent a minority whose influence transcends its size or its political and material power. As always, the rebellion embraces elements that are timeless, and other elements that reflect the great issues of the day. And, as always, rebellion is not a unitary process. Rather, there are several rebellions going on at once. Three of them are here called the Red, the Black, and the Green.

The Red Rebellion refers to the New Left. The movement is labelled "red" only because of the halo effect from the traditional color of bolshevism and the internationale as seen by our generation. Actually, the voice of today's underground press reveals that the New Left has little use for ancient organizations like the American Communist Party (most of whose members are over 30 if not over 60). However, since their foes inevitably call them red, they may adopt the color out of sheer defiance; that we assign it to them here is less political euphemism than

---

Based on studies supported in part by a grant from the Foundation Fund for Research in Psychiatry, Inc., by the senior author's Fellowship Award at the Center for Advanced Study in the Behavioral Sciences, Stanford, California, and by NIMH Senior Stipend 1-F3-MH-35063-01.

poetic license. Even fairly recent definitions of the New Left have included in it (or identified it with) Black Power groups, and hippie elements have not been excluded.

Jack Newfield, in the preface to his book, *A Prophetic Minority,* says: "I define the New Left broadly to include organizations like Students for a Democratic Society (SDS) and the Student Nonviolent Coordinating Committee (SNCC); *ad hoc* decentralized movements like the Berkeley Free Speech Movement (FSM) and the movement against the war in Vietnam. It includes idealistic Peace Corps and VISTA (Domestic Peace Corps) volunteers and nihilistic Berkeley bohemians; new institutions like the Institute for Policy Studies and the Mississippi Freedom Democratic Party; new publications like *Liberation* and the *Southern Courier*; individuals like Bob Parris, Tom Hayden, and Staughton Lynd; it even spills over into sociocultural movements represented by Bob Dylan, Phil Ochs, Dr. Timothy Leary, and Allen Ginsburg."[1]

From another point of view, Phillip Abbot Luce, author of *The New Left,* cites the democratic socialist and editor of *Dissent,* Irving Howe, as saying: "There is a new radical mood in limited sectors of American society . . . For this disturbing minority I have no simple name; sometimes it looks like kamikaze radicalism, sometimes like white Malcolmism, sometimes like black Maoism."[2] Luce, though, does have a simple name: "The name that has gained general acceptance when describing the 'new radical mood' is The New Left."[2] Luce's definition of the New Left comprises such organizations as the Progressive Labor Party (PLP); the W. E. B. DuBois Clubs; the Trotskyist Youth Socialist Alliance (YSA); splinter groups such as the Spartacists and the American Committee for the Fourth International; the Workers' World Party and its youth arm, Youth Against War and Fascism; and such wild-eyed militants as the Provisional Organizing Committee, the Committee to Aid the National Liberation Front of South Vietnam, and the Revolutionary Action Movement.[2]

However, Newfield distinguishes the New Left from what he calls the Hereditary Left, represented by the Progressive Labor Party, the W. E. B. DuBois Clubs, and the recently dissolved May 2nd Movement. He contends that "Despite surface similarities with the New Left, the Hereditary Left is actually an ideological extension of the old 1930's left, is ideologically Leninist in structure and outlook, and is oriented toward either China or the Soviet Union rather than forging a new vision of American society."[1]

Paul Jacobs and Saul Landau, in their book *The New Radicals*, feel that "to be in The Movement is to search for a psychic community, in which one's own identity can be defined, social and personal relationships based on love can be established and can grow, unfettered by the cramping pressures of the careers and life styles so characteristic of America today."[3] They state further:

> "The new movement is also a revolt against the postwar 'over developed society,' with its large bureaucracies in government, corporations, trade unions, and universities. To those in the Movement the new technologies of automation and cybernation, with their computers and memory bank machines, are instruments of alienation, depersonalizing human relations to a frightening degree . . . Technology's emphasis on routine efficiency has created a set of values, rationalized by its supporters as representing 'the facts of modern life . . .' The decision-making over which the governed no longer have control extends far beyond politics: in the technological order every aspect of the people's lives is under the control of administrators far removed from responsibility to the governed."
>
> "Above all," they continue, "Those in The Movement now restlessly seek to find a new politics and a new ideology that will permit them to link existential humanism with morally acceptable modes of achieving radical social change."
>
> "The Movement feel that modern American liberals have substituted empty rhetoric for significant context, obscured the principles of justice by administrative bureaucracy, sacrificed human values for efficiency, and hypocritically justified a brutal attempt to establish American hegemony over the world with sterile anti-Communism."
>
> "What began perhaps as a rebellion against affluence and liberal hypocrisy grew in a few years into a radical activism that protested injustice at the very core of the society. But when even this was tolerated by the structures that were under attack, some of the young radicals began to think about something beyond rebellion or radical protest. The Movement now is struggling to develop an ideology that will guide them toward building an organization that can compete for political power."[3]

Irving Howe, viewing the movement from another vantage point, sees as "characteristic attitudes among the 'new-leftists':[4]

(1) An extreme, sometimes unwarranted, hostility toward liberalism,
(2) An impatience with the problems that concerned an older generation of radicals,
(3) A vicarious indulgence in violence, often merely theoretical and thereby all the more irresponsible,

(4) An unconsidered enmity toward something vaguely called the Establishment,
(5) An equally unreflective belief in 'the decline of the West,'
(6) A crude, unqualified anti-Americanism . . .
(7) An increasing identification with that sector of the 'third world' in which 'radical' nationalism and Communist authoritarianism merge."

In contrast, M. Stanton Evans offers the simplest, and perhaps most comfortable view: "On all doctrinal matters, the New Left is merely American Liberalism writ large, demanding that the regnant abstractions be enforced *instanter* and across the board, with no exceptions . . ."[5] And not long ago Andrew Kopkind took the most inclusive position: "The hippies, the poor blacks, and Stokely Carmichael are, in varying degrees of attachment, elements of the New Left, and they are trying to say something that will be taken seriously—not about themselves, but about the world in which they are uneasily placed."[6]

Recently, however, the distinctiveness of the three rebellions has become increasingly apparent. The Black Rebellion more and more rejects the other two, for reasons that range from "reverse racism," anti-Semitism, and reaction formations against identification with the aggressor, to conscious political opportunism among certain leadership elements, and honest despair among others whose hopes of real help from the white community have been disappointed again and again.

Today, The Black Rebellion refers to what the community at large discerns as the Negro Revolt in its more strenuous manifestations. Actually it is a sweeping change in the orientation and behavior of the Negro younger generation, representing as much a rebellion against the compromises and traditions of the Negro older generation as against the white world. Here no poetry is required. The movement calls itself Black with great pride.

The Hippies make up The Green Rebellion. The color, from our brush, symbolizes their love of nature ("flower children"), verdantly ingenuous ideals, and, of course, "grass."

The hippies are something new, and something else. The Green Rebellion is a fast growing new species of hybrid flowering plant, with shallow but numerous roots among the older Bohemias, the beatniks, the surf set, the shadow campuses, the folk rock followers, the God-is-dead mourners, the school dropouts, college dropouts, church dropouts, establishment dropouts, and the lotus-eaters—users of speed (amphetamines), downers (barbiturates), peyote (mescaline), DMT (dimethyltryptamine), STP (a phenylethylamine with additives), acid (lysergic acid diethyl amide) hash (hashish) and pot (marijuana).

The upsurge of hippiedom achieved sufficient direction, identity, and phenomenological consistency to be termed a rebellion about three years ago. The "hippie" label was then just beginning to be differentially applied. There were only a few gathering-places: New York City's lower east side slum neighborhood called The East Village (because of its latitudinal relationship to the older bohemia of Greenwich Village); the Sunset Strip area of Los Angeles with its special nonalcoholic hangouts for teen-agers; and the Haight-Ashbury neighborhood of San Francisco. In the summer and fall of 1966 it was "The Hashbury" that emerged as the fountainhead of hippiedom. In twelve short months, The Green Rebellion spread like the pollen of its own grass, borne on the bosom of the prevailing westerly winds from San Francisco Bay across the land and even across the Atlantic, and fertilized by the ubiquitous media, hot and cool.

The similarities among the three rebellions are superficial; the differences run deep. The Red Rebellion is political, theoretical, intellectual and radical. The Black Rebellion is economic, social, racial and activist. The Green Rebellion is cultural, religious, mystical, and passively pharmacological.*

The participants in all three groups affect costumes quite different from the accepted garb of their parents. They wear old clothes, quaint garments, colorful motley, or the raiment of far-off lands and times; battered sandals, outlandish boots, or nothing at all on their feet. Some sport strange decorations, cabalistic artifacts, beads, bells, pins, labels, signs and badges with messages bearing many meanings. And withal they favor a hirsute atavism that suggests at best the elegantly bearded Renaissance, at most beatific the unbarbered Bible, and at worst the hairy Pleistocene.

All three Rebellions scorn the appurtenances of affluence but for different reasons. The red rebels seem to pledge sartorial allegiance to an historical and political symbol: the masses of the proletariat (now more

---

*Even fairly recent excellent discussions of "the youth scene" have not documented the differentiation of the hippies from college campus groups.[7] Good current reviews of hippiedom are available, however, including von Hoffmann's series[8] and the Time compendium.[9]

often simply called the poor and underprivileged); button-down collars are for the bourgeois. Some black rebels wear old clothes because that's all they possess; others, who enjoy better economic fortune, wear overalls as a socioeconomic symbol of racial solidarity. But the costumes of the green rebels are symbolically spiritual: old, for the rejection of materialism and of selfish affluence; colorful, for the joy of psychedelic self realization; and exotically trans-cultural for the oneness of mankind and the appreciation of truths and insights to be found in a variety of religions all around the world.

All three rebellions oppose the military involvement of the United States in Vietnam. The reds are against it for historically political reasons ("Imperialism, colonialism!"). The blacks are against it for socioeconomic reasons ("A white racist war that squanders resources needed to correct ghetto conditions at home!"). The greens are against it, as they are against all strife and violence, for spiritual reasons ("Love is what's happening, man; like, its nowhere to go around hating people when we're all part of each other!").

Rebels of all three classifications take a generally dim view of modern psychiatry and psychoanalysis. The New Left thinkers tend to look upon the discipline as a quaint and perhaps fraudulent if not downright reactionary branch of establishment medicine, which is generally seen as a profession politically somewhat to the right of banking. Psychotherapy is for the rich, and psychoanalytic theory is a formulary of obvious truisms and middle-class mythology deriving from a hypocritical pseudocommitment to nineteenth century morality. The existentialist orientation is by and large much more compatible with the New Left position than are the cause-and-effect formulations of ego psychology.

The Negro revolutionaries are deeply suspicious of psychiatry, seeing it not only as neglectful of the poor, black, and underprivileged, but also as part of a general white (particularly Jewish) conspiracy to deflect the new black aggressive propensities. "Sure, Dr. Goldberg says we're maladjusted; he says work on your inner conflicts and learn how to adapt to your circumstances. That'll be fifty dollars, boy. Don't they think we see through that crap? It's the Jesus bit all over again, only more expensive."

The hippies (typically more tolerant) see the psychiatrist as a nice man trying to do his thing, but under hopeless conditions: "Man, most of these shrinks have hang-ups worse than anybody's. Then they try to talk some cat into being happy though straight. Why, that's impossible!

And, like, they give them the wrong dope (drugs) until they are like zombies. They ought to blow their own minds first, man, and then turn on with their patient and together they might find out where it's at." The mistrust of hippies for establishment medicine (especially psychiatry) was one of the major complications in the operation of the Haight-Ashbury Free Medical Clinic; the more it grew to offer something resembling respectable standards of medical care, the less inclined hippie-types were to make use of it.

When the senior author went to the San Francisco Bay area on a year's fellowship at The Center for Advanced Study in the Behavioral Sciences, Stanford, California, beginning September 1966, one of his first extracurricular activities involved going into the Haight-Ashbury district to find a fifteen year old runaway boy, the son of a colleague. Living in a hippie pad, the lad was already deeply identified with the movement. This and subsequent expeditions to the Haight-Ashbury led to a growing interest in the remarkable subculture that was rapidly developing there, and in the months that followed a number of phenomena were observed at first hand, including the following.

The Diggers, (an anonymous group of service-motivated hippies primarily concerned with feeding the hungry) was established. A hippie organization called The Switchboard came into being as a communications center to enable outsiders to contact people living in the Haight-Ashbury who could not otherwise be located. Church groups organized a special home, Huckleberry's for Runaways. The HIP Job Corps, a hippie-run employment agency, was organized to make it easier for them and others in the Haight-Ashbury district to find paid work that might suit them. The Flame was, for a time, a sort of philosophical "city hall" of the hippies.

A series of extraordinary hippie-sponsored gatherings began with the First Human Be-In at the Golden Gate Park in January 1967. This and its successors were characterized by widespread expressions of love and brotherhood, the sharing of appreciation for the beautiful setting, the music (often provided free by hippie-sympathizing rock and roll groups), and the gleeful delight of smoking marijuana in the company of thousands doing the same thing under circumstances that virtually ensured safety from police harrassment or arrest. Such happenings were well publicized, by the "straight" press as well as by such "underground" publications as The Berkeley Barb and, of course, The Oracle which is the unofficial voice of the Haight-Ashbury.

The Free Medical Clinic was organized and for a time largely staffed by hippies and hippie sympathizers aware of the high rate of illness in the Haight-Ashbury. Venereal disease, hepatitis, malnutrition, bronchitis, and drug reactions were endemic, while the sufferers—especially newcomers—often avoided the usual medical channels for fear of bringing their illegal use of drugs to the attention of the authorities.

By June of 1967, it was clear that more detailed observations would be needed if the hippie subculture were to be studied in depth. A semipermanent observation post within the Haight-Ashbury district seemed desirable. Therefore a typical large apartment, or "pad," was obtained, cleaned, disinfected, humbly but suitably furnished, and decorated with posters, flowers, and paint. For the next six months an interdisciplinary study of the hippies was undertaken.

At full strength during the summer months, when a substantial influx of young people (some 30,000 from all over the country) inundated the Haight-Ashbury, the team consisted of a program director (L.J.W.), a child psychiatrist and co-director who was on the scene full time (J.R.A.), a graduate student in cultural anthropology (Reed College and Oxford), a graduate in psychology (Stanford), a recent graduate in humanities (Goddard), a premedical student (Brandeis), an undergraduate in photography and graphic arts (San Francisco State College), and an undergraduate in behavioral sciences (Stanford).

The group addressed itself particularly to (1) the structure of the hippie subculture; (2) patterns of drug use and abuse; (3) acute drug intoxications (delirious reactions, dissociative reactions, "bad trips", etc.); (4) long term complications of drug use including subtle personality changes, habit deterioration, and alterations of intellectual capability; (5) teen-age runaways and "teeny boppers"; (6) sexual problems including promiscuity, venereal disease, perversions, and illegitimacy; and (7) intergroup conflict and violence. This work involved the cooperation of health agencies, the chief of police, the fire commissioner, the housing commissioner, and social and behavioral scientists from other institutions including Stanford University, the University of California Medical Center, and Mount Zion Hospital in San Francisco.

The Haight-Ashbury district proved to be a neighborhood in flux, once elite, now considerably run-down, and comprising about 70 per cent Negro inhabitants. Landlords had been functioning more and more on an absentee basis and neglecting their property. As the desirability of the neighborhood for white middle-class dwellers diminished, relatively large houses (more or less dilapidated) became available at reasonable rents.

The first hippies moving into this area were a small group of people (not all of them young) who were deeply philosophical or in pursuit of various eccentric religious ideals. Some had evolved out of the beatnik group of San Francisco's North Beach which, like Greenwich Village, had become fatally fascinating. (Tourists quickly grew to consider North Beach a must. Hard on the heels of the tourists came the petty crooks, hawkers, pitch-men, prostitutes, topless entertainers, bars, dance halls, restaurants, and other commercial enterprises, all of which made North Beach progressively less desirable for the bearded thinkers, their friends, and their families. The same process is already at work on Haight Street.)

The Negroes living in the Haight-Ashbury district were initially surprised, then amused, then annoyed, and finally deeply alarmed and concerned with the rapid hippie influx. It was hard for them to understand why a group of middle-class and upper-class white young men and women should be drawn to their shabby neighborhood. The threat to the Negroes increased as the ranks of the hippies swelled with converts, sympathizers from nearby San Francisco State College, and a growing number of Bay area dropouts of all ages. There was a greater competition for housing. The colorful world of psychedelphia was turning "The Haight" into a round-the-clock street carnival. Tourists and gawkers created traffic jams. Prices began to rise. And the flamboyant use of illegal drugs, particularly marijuana and lysergic acid diethylamide (LSD), attracted a steady increase of police activity.

Racially mixed couples in the neighborhood (mostly Negro men with white wives) found that the precarious stability of their situation was being shaken by the easy and loving ways of the hippies, with whom these couples in fact had nothing whatsoever in common. But it was the Negro youth, especially those participating in the Black Rebellion, who became particularly antagonistic toward the hippies. These young Negroes were espousing a black racism based, in part, on their own stereotyped image of the white establishment. Then came the hippies, who didn't fit such stereotypes, and who greeted suspiciousness and rejection by the young Negroes with insufferable loving-kindness.

The Negroes did not view the willingness by hippies to live in their neighborhood as a sign of egalitarian acceptance. As one young militant

Negro leader put it, "These cats come down here and live like they are poor—maybe even panhandle on the streets—and act like they've got less than we have; but man, any time they get tired of it they can shave and shower and hop a bus and before you know it there is 'welcome home, baby' out in the lily-white suburbs, and then they can talk about how they lived for a while in a slum with the niggers."

Most important, perhaps, is the conflict in direction of movement between the two rebellions, Black and Green. The blacks see a white man's world of affluence and material possessions that are the *desiderata* of the culture. Basically they accept the values of the culture. What they want is their fair share of it, an opportunity to become truly a part of it. They want to get out of the Haight-Ashbury and into Woodside, Mill Valley, Russian Hill, Seacliff. But along come the hippies, in the opposite direction, saying, "Like, man, we've been there, and it's nowhere."

Here then are whites who are turning their backs on the very things the Negroes are striving to obtain. This poses a deeper threat to the black rebels than does white racial prejudice, because it allies these strange whites with many of the attitudes of the black rebels' parental generation—a gentle, patient, loving, accepting, and basically religious adjustment to uncomfortable and deprived circumstances.

It is primarily on this basis that there are practically no Negro hippies, although many Negro boys prowl The Haight seeking hithertofore unavailable and taboo sexual opportunities with white girls. But even such liaisons, often initiated as "one way to screw the whites," can be troublesome for the black rebels. A Negro boy from Chicago said, "All these spade cats looking for white chicks depressed me, until I met this hippie chick and she just flashed on me and took me over and turned me on. It blew my mind, man, it was great, but now I've got to split; because if I hang around her and her friends I won't be able to tell who whitey is any more. And if I don't know who's an ofay, I may not even know who I am myself." The rare Negro who turns hippie nearly always proves to be from black bourgeoisie in Cleveland or New Rochelle or Beverly Hills—not from Harlem, Hunters Point, or the Haight-Ashbury.

The lower class Negro family, struggling to rise into the middle class, has viewed this peculiar white influx into the Haight-Ashbury as a homogeneous movement. The hippie-types all look alike to them. Many newspapermen, police officers, narcotic agents, underworld operators, establishment commentators, and even behavioral scientists have made the same mistake. While it has been estimated that there are 25,-50,000 hippies in the Bay area, 10,000 is probably a more realistic guess, with perhaps half this number in or near the Haight-Ashbury district. Only a fraction even of this group should be considered "true" or "hard core" hippies. Others, often mistaken for hippies, have come to San Francisco searching for excitement, mystic revelation, free sex, escape, acceptance, drugs, or even sanity in what is for some a therapeutic community of sorts. Most of them are youngsters who dress and talk like hippies, and use drugs, but themselves realize the difference. They will say, sadly, "I'd like to be hippie but I can't make it," or "I'm not good enough," or "I'm really just a speed freak," or "I've got too many hang-ups to really make the scene; when I turn on I freak-out everybody else; the hippies just put up with me 'cause they're so cool."

What is a true hippie? Unlike participants in the Red and the Black Rebellions he is not an activist. The New Left and "active" pacifists like singer Joan Baez, are openly critical of the hippie for his passivity. Rather than to war against the establishment, or to struggle to change it, he has dropped out of society, turned his back on it, dissociated himself from it, and is attempting to create an entirely new way of life. He is peaceful, nonviolent, and committed to an ideal of universal brotherhood, love and sharing. His expectation is that in time more and more people will see the wisdom of his course and follow it for themselves. His background is most likely to be white middle class, although upper class and upwardly mobile lower class families are also represented. A survey by the United Jewish Community Centers in San Francisco in November, 1967, found that between 20 and 25 per cent of the Haight-Ashbury hippies were Jews.[10]

While many hippies are highly intelligent, it is not an intellectual movement. Basically it is a spiritual movement. A sense of oneness exists among the members of a hippie commune or tribe. This communion derives from the mutual experience of revelation through psychedelic delirium. At the same time, strong permanent interpersonal commitments are few; nobody wants to relinquish personal freedom or mobility. Repeated episodes of LSD intoxication ("trips") have induced in these people sensations of transformation, insight, oneness with all living things and beatitude. There are other intoxicants (opium for instance) that can effect similar subjective experiences, but such drugs obtund consciousness at the same time. LSD paradoxically heightens awareness while impairing the sensorium. It affects the nervous system in such a way that information processing seems to be increased; at any rate there is increased awareness of it. The well established automatic screening processes that prevent us normally from being flooded by sensations from without, or swamped by excessive awareness of the ongoing information processing within our own brains, is substantially altered. At the height of the LSD reaction the net result is a "jamming of the circuits" resulting from "information input overload," accompanied by a sense of extraordinary contact with both the world within and the world around. Ideas then tumble over each other chaotically while intense feelings dominate the sensibilities. When the surroundings include other people going through the same experience, the characteristic sense of interpersonal communion and sharing occurs. Everyone is tuned in to his own primary process thinking which can't be described, but can be understood empathically by others who are similarly "high," "stoned," or "tripped out."

Virtually all of the true hippies have had this experience repeatedly, many feeling that through if they have reached a state of mind that constitutes complete self awareness so that no further trips are needed. Some show personality changes that, to the trained observer, may seem schizoid, or similar to certain organic brain syndromes (e.g. postlobotomy) that leave the individual more comfortable but less able to carry out complex long-term plans, endure frustration, concentrate for extended periods, follow routines, or successfully master new material (learning) with the same ease as before.

The group LSD trip is approximated by ceremonial group intoxications with other drugs such as mescaline, bufotenine, or psylocibin. The fellowship of drinking companions is not of the same order. More closely related is the camaraderie of marijuana smokers.

To a large extent the smoking of marijuana is the cement that holds the hippies together, along with the social dynamics of guiltless lawbreaking. One might call Haight-Ashbury "The Asphalt Sherwood Forest" because of this state of affairs; here are the merrie men, gathered together under the greenwood tree (Golden Gate Park); outlaws all. In California possession of LSD is a misdemeanor, or marijuana a felony, but one 15 year old lad blithely confessed (when "busted" by "the fuzz" and found to be "holding" marijuana), "Oh, I don't smoke pot myself, I just carry it for a status symbol!"

The current proponents and opponents of marijuana are severely handicapped by lack of sufficient information about the long-term effect of tetrahydrocannabinol (the principal active component of cannabis) upon the brain. The next few years will bring a flood of new data about marijuana (pot, grass, weed, boo, reefers, Mary Jane, kif, bhang, hashish, etc.). Good advice to youngsters today might be, "Just because alcohol is bad, marijuana isn't necessarily good; just because the present law is absurd, marijuana shouldn't necessarily be uncontrolled by law; just because you've been lied to about its dangers, marijuana isn't necessarily harmless." But it is doubtful whether dire warnings or ominous laboratory findings would deter many of the hippies from their pot-smoking ways.

If marijuana is the glue of hippiedom, LSD is its sacrament. The mind-fragmenting nature of LSD intoxication contributes to the subsequent sense of being reborn after having been transported, as Aldous Huxley put it, on a trip to "The Antipodes of the mind;" it commits hippies to a common mystique, a sense of mutual revelation. For the true hippie, like Saul of Tarsus who saw the light on the road to Damascus and became St. Paul, the transformation is global and often results in dedication to a new way of life.

In the pads along Telegraph Avenue in Berkeley, where many of the New Left students and shadow campus types collect, a superficial resemblance to the hippies may be seen. Old clothes, beards, beads, buttons, and bare feet abound. The aroma of marijuana can be discerned in the ambient air. But in the pads there is a ferment of talk, quotation and counter quotation, disputations *ad hominem,* and generallly noisy but clearly intellectual uproar. The walls are lined with books; background music is often ignored; when attended it can be heard to range from Bach to Brubeck to Baez.

The contrast of this to the typical Haight-Ashbury pad is substantial. Because his road to revelation was perceptual and emotional rather than verbal and intellectual, the hippie tends to reject semantic approaches to the discovery of truth. Thus hippies keep very few books. Many of them are for laughs (e.g., comic books). Volumes on prophecy and astrology are often found in their pads. Large numbers (probably a majority) of hippies believe in one or another manifestation of magic, ESP, reincarnation, astrological prophecy, and oriental auguring of the "I Ching" variety.

Music in the hippie pad is often deafening and precludes conversation. Heavily favored are "acid rock," group recordings by such hippie favorites such as The Jefferson Airplane and the Grateful Dead, and oriental renderings on exotic instruments like the koto or the sitar.

Hippies engage very little in argumentation or protracted discourse on a single topic. In fact, there often seems to be an attempt to communicate the most possible in the fewest words, and even to restrict the vocabulary so that a small number of words and phrases (constituting a characteristic argot or patois) will suffice both to transmit whatever digital information is necessary in the form of words, and at the same time to constitute recognition of the fact that analog communication of affect is of much greater significance and value.

Thus the hippie is unlikely to formulate his credo in words. If he did, it might be something like the following:

"We believe that the entire development of human society up to this point represents a one-way road to catastrophe. We think that the only hope for humankind is to go back to Nature and start over. The world has gone crazy with selfishness, with armaments, with violent destruction of individuals by ones and twos and millions."

"We see no point in storing up all the material goods which everybody seems so viciously to covet. The terrible competition among and within the major societies of the world today is not only unhealthy but is so destructive that it is going to bring down what little civilization exists."

"The values that are exemplified by the establishments of all the great powers are abhorrent to us. The very use of the word "power" describes what is wrong with the whole system. We reject the formal structures that exist within the establishment, ranging from assembly-line educational institutions to all of the organized forms of religion, which as soon as they build themselves a church have already defined themselves out of the spiritual and into the material world."

"What we seek is the opportunity for human beings to find themselves as individuals among Polynesian natives, African tribes, Oriental cultures, and, more and more, American Indians. There are potentialities in each individual that he may not even suspect, automated as he is by all the different systems that our society presses upon him and pulls him through. He must get out of all this ponderous regimentation, be liberated from all of these sterile and destructive values, and be given a chance for the first time in his life to become himself. Let him find out (through drugs) the possibilities that lie within his own mind, and actually do day by day that which he enjoys, is really best suited for, and finds fulfilling."

"We believe that people should live at peace, share freely with each other, dwell in harmony, appreciate the bounties of nature, and, being closer to it, become more respectful of it. Not poisoning the food with chemicals! Not contaminating the waters! Not polluting the very air!

"In the love of nature, of life, and of each other, seeking the ultimate of freedom, individuality, self-fulfillment, and generosity, let us then live together with others under conditions that make it possible for us to give fully of ourselves and thus achieve total self realization."

That would be a long speech. No hippie would ever make such a speech. If he were talking to other hippies, he would transmit in a few words the entire meaning of it. He would say, "Man, like everybody just ought to do their own thing." The listeners would murmur, "Yeah!" This is the hippie credo.

How does a hippie achieve his identity if learning through verbal transaction is limited? The philosophy apparently evolves with the individual's passage through certain stages that might be listed as follows:

(1) Despair or disgust with the world as it is today (often identified with one's own middle class parents and their way of life).

(2) A search for meaning, and for self, in the light of a good educational and cultural background, and from an initial posture of financial (not interpersonal) security.

(3) Association with other searchers, some of whom seem to have discovered a way.

(4) Enhancement of camaraderie with these fellow-pilgrims by marijuana smoking which (a) usually generates a state of closeness through heightened awareness and sensory stimulation (followed, of course, by lethargy) and (b) creates a cabal outside the law, with all of the dynamics of a persecuted minority group coming into play. Dr. Timothy Leary would call this phase "tunning in."

(5) "Turning on." While this term is used to describe many types of stimulation, "blowing the mind" with LSD is the ultimate "turn on" experience. Its mind-expanding sensations may lead to a feeling of cosmic consciousness, often followed by a "re-

birth" including the characteristic sense of oneness with mankind and universal love.

(6) "Tuning in." Here the new hippies really being to appreciate the mutuality of their experiences. Often while coming down from highs or trips they "rap" to each other (i.e. converse excitedly while strongly *en rapport,* revealing and sharing all manner of thoughts, feelings, and personal history). Quickly, by observation and word of mouth, they learn ("tune in on") what's happening—the true nature of "the scene". Most hippies today say that you can't really "tune in" until after you've been "turned on".

(7) "Dropping out." This implies a major break with "the straight world," followed by the assumption of a new way of life and a new set of values.

It is of interest that Leary, who formulated "Tune in, turn on, drop out!" and is often blamed for what transpired subsequently, was using terms from the new slang to describe what was already happening. His influence among the hippies today is much diminished. Their anarchistic tendencies make them deeply suspicious of organizers, generals, governors, and messiahs. Dr. Leary is usually dismissed rather casually: "That cat was really with it once; but, man, now he's on this big ego trip all the time." The more thoughtful (including some who now eschew the use of LSD) see him as a tragic figure, a victim of his own self experimentation and of a self-righteous establishment lacking in compassion.

Once the hippie has "dropped out," he is likely to enter into some sort of nondemanding, loosely organized liaison with others in a "commune," or "tribe". There is great variety among these groups and their pads. Some are dirty, others clean; some are affluent, others broke; some are weird, others actually mimic the straight world. By and large, the more completely the pad is able to function in keeping with the unspoken credo, the more admirable or "cool" it is considered to be. Ironically, the coolest pads may come to resemble upper middle-class apartments that are avant garde enough to espouse the new "psychedelic" art forms and costumes.

Sexual practices among the hippies are strictly up to the individuals involved. They vary from monogamous marriages to genuine group marriages. The promiscuous and the virginal are both accepted without question. Privacy is a matter of taste. There is a definite trend toward nudism when conditions permit.

Work habits are highly individualized. If a pad is really cool there will always be money coming in from one source or another enough for the essentials of life such as spaghetti, dope (the hippie's term for all desirable drugs), milk, and matches. Chicks and cats drift in and out, bringing and sending greetings, seeking pleasure in listening to music, walking in the park, dancing, strolling on Haight Street, rapping to their friends. For hippies work is each person's thing, from part time house-painting or carrying the mail to regular employment (teaching, performing, handicrafts, graphic arts, etc.) Work is acceptable if it's not a sellout—if it's truly enjoyable—if everyone is in fact "doing his own thing."

Until recently true hippies have been able to survive and find a measure of contentment in places like the East Village or The Haight. This capability may be diminishing. The hippies are increasingly menaced by forces emanating from the establishment (e.g. police, narcotics officers, etc.); the predatory underworld (preying constantly upon the gentle and the defenseless); other more aggressive rebel groups (e.g. the motorcycle gangs); and perhaps most of all from the onslaught of pilgrims. The majority of these newcomers are not what might be called bona fide hippie material, but toward them the hippie obligations of kindliness, sharing, and acceptance must be manifested. The newcomers swamp the hippies. They crash (sleep) in their pads, consume their food, plead, "Lay some bread on me" (give me some money) when they are broke, attract the police (who are always looking for runaways,) have bad trips on good dope, and bum-trip other people (i.e. upset them while they are particularly vulnerable during a trip) out of ignorance of proper procedure. Furthermore the numerous would-be hippies, pseudohippies, plastic-hippies, and teeny boppers are highly vulnerable to disease, attack, arrest, and other forms of disaster.

Before the great influx during the summer of 1967, life was already sufficiently hard in the Haight-Ashbury to challenge all the hippies' resources of "Flower power": beautiful philosophy, denial of reality, and chemical self treatment of untoward emotions with various drugs. Now, the fuzz is cracking down, the teeny boppers are "shooting speed" (dangerous intravenous injections of methamphetamine), dope is harder to come by and the Mafia are moving in on the amateur sources of supply. A rougher white element is starting to take over the hangouts on Haight Street, and is formenting trouble with the already truculent Negroes.

Under these pressures, singly and groups, many hippies are resettling themselves along the west coast, in villages, farms, parks, woods, even caves. New hippie-style enclaves are springing up all across the country, although it is not possible to say just how many are deserving of the generic label. Even in Europe one may now find growing numbers of American-style hippies (who must be distinguished from the more activist New Left-type "Provos," also on the increase.)

One of the most interesting things about the hippie scene has been the reaction of our society to it. Some people have been violently offended by the very appearance of hippies. It is amazing, with all we have learned about human behavior, and all that has transpired in the development of modern educational institutions, that so many school administrators permit themselves to be drawn into ferocious confrontations with students on the subject of haircuts. It is as though the risk of hippiedom intrudes along with the styles thereof. This risk is seen by many citizens as the vanguard of evil, the quintessence of all things parasitic and degenerate in our society, and the hippies are viewed as an infiltrating tribe of useless filthy gypsies who should not be permitted to remain in any respectable community.

Other observers have taken a completely opposite viewpoint, seeing the hippies either as a harmless and amusing combination of fads and freaks, or (more seriously) as a joyous and even saintly band of spiritual truth seekers and karma storers. Bishop Pike has compared them to the early Christians. Toynbee sees their emergence as a warning to Western civilization. Some suggest that the last best hope for survival of the human race may lie in the Green Rebellion and its sublimely simple return to Nature. This romantic view, no less than the paranoid view, reflects the degree to which people project onto the hippies their own hopes, fears, preoccupations and unconscious wishes.

Certainly one cannot help wondering whether those who seem to be so threatened by the hippies may not in fact be struggling with whatever is stirred up inside themselves through stimulation of fantasies of freedom from responsibility, escape from the frustrations of rigid life situations, chemicals that will provide both surcease from unpleasant emotions and miraculous generation of unbounded libidinous resources, and uninhibited sexuality without the necessity for interpersonal commitments. The hippies sense this. Many of their antics are elaborate jokes on the establishment.

In the light of our studies, the hippies can be seen in another way. Here is a group of people, obviously sensitive to aspects of our civilization that are so dreadful that most of us avoid thinking about them as much as possible: war, violence, cutthroat competition, grasping materialism, sexual hypocrisy, interpersonal isolation, pollution of the elements, loss of contact with nature, loss of individual identity in the massive treadmills of the modern megalopolis. The hippies (ugly, dirty, or grotesque though they may seem) seek beauty, freedom, creativity, individuality, self expression, mutual respect, and the ascendence of spiritual over material values.

These goals and aspirations are clearly derived from the traditions of great religions and ethical systems. The ineffable tragedy of the hippies is that their Green Rebellion, while still growing, is doomed to fail. It is tragic that individuals capable of formulating such lofty ideals are being driven to self intoxication with powerful drugs in order to imagine their fulfillment in 1968. The very chemicals they use will inevitably enervate them as individuals, and sooner or later will bleed away the vitality of the hippie movement to its death.

REFERENCES

1. Newfield, J.: A Prophetic Minority. New York, The New American Library, 1966.
2. Luce, P. A.: The New Left. New York, McKay, 1966.
3. Jacobs, P., and Landau, P.: The New Radicals. New York, Random House, 1966.
4. Howe, I.: New Styles in "Leftism". In Steady Work. New York, Harcourt, Brace, & World, 1966.
5. Evans, M. S.: Orthodox Rebels (A review of "The New Left"). National Review, July 12, 1966.
6. Kopkind, A.: Taking Radicals Seriously (A review of "A Prophetic Minority"). New Republic, December 10, 1966.
7. Simmons, J. L., and Winograd, B.: It's Happening. Santa Barbara, California, McNally Publishing Co., 1966.
8. Von Hoffmann, N.: Series of articles on the hippies, Washington Post, Fall, 1967.
9. Brown, J. D. (Ed.): The Hippies. Time, Inc., New York, 1967.
10. Problems Involving U.S. Jews. San Francisco Chronicle, Thursday, November 30, 1967, p. 35.

Discussion by Earl G. Witenberg, M.D.

The heuristic classification of the Rebellions as Red, Green and Black is powerful and useful. The authors evidently feel the Red and the Black are easier to understand—they fit into the framework of traditional rationalistic rebellions. They represent the timeless and current issues for which most modern rebellions have been fought. The implication is that they will eventually be integrated into the mainstream to a greater or lesser degree. They will lose their identities in the typical American way of accommodation. The interrelationship of the Black and Green is as discussed.

Dr. West pays most of his attention to the Green Rebellion—"cultural, mystical, religious, and (of all things) pharmacological". The hippies are a "fast-growing new species of hybrid plant". The roots are in "the older bohemians, the beatniks, the surf set, the shadow campuses, the folk-rock followers, the God-is-dead mourners, the school and college drop-outs and establishment drop-outs, and the lotus-eaters". Where are the families as part of the roots or signal systems for this behavior?

It's important to us as psychoanalysts to understand the family scene that makes this rebellion necessary. The hippies are people who were "in" and who chose to "drop-out"—presumably to find themselves. "The first hippies were deeply philosophical and pursued eccentric religious ideals." I am not clear as to the meaning of this. Are these eccentric ideals the ideals of peace, nonviolence, universal brotherhood, and sharing? Twenty centuries ago there was a man who lived these ideals. Was he eccentric? Maybe. Eccentric in this culture? Definitely yes. Could it be that the culture is eccentric? The authors are wistful and sad that the seeds of its destruction are intrinsic to the movement. That it is driven out of desperation to using the harmful drugs to imagine it has attained the lofty ideals it has set.

What makes it possible for the true hippies to believe that drugs—marijuana and/or LSD—increase their own perceptions of themselves and the real world? Have not other people over the years done the same thing without drugs? Throughout the known history of mankind, there have been people who have had to retreat from civilization to sharpen their perceptions of what is real and unreal, what is meaningful and what is meaningless. Some have become hermits. Some form communities. We all do it in one way or another. We bird watch, hike, meditate. We go into analysis to expand our consciousness. We take a trip by auto or by plane or by regressing temporarily. We have weekend homes, summer homes, summer vacations. We move from a city to a suburb, to an small island.

What makes for the belief in the chemical trip? Have the hippies unwittingly carried into their culture our faith in technology, in this instance, pharmacology? We participate in the barbiturate for sleep, the dexedrine for diet, the tranquilizer for anxiety culture. Some of us make a living this way. So the hippies have learned their lesson well, and in this drug way, caricature the belief of us elders. The roots in other religions using drugs are clear.

They also seem to caricature the "togetherness" which was so preached as good child-rearing practice for their generation. They get together by having the common vision as to the true nature of the scene. They turn on and tune in but only on one channel.

They lay bare and mock our great discoveries—the affluent society, the A-Bomb, the computer, the importance of intellectual achievement, our communication-minded society, our chemistry.

They have few strong one-to-one, I-thou relatedness because of their fear of loss-of-freedom and motility. They practice individualism in a way that could warm the heart of a William Buckley. In their child-centered upbringing, wasn't there an abdication of responsibility by some parents? Didn't they sense a promise that the world could be as they wanted it to be? Isn't that the source of their disillusionment and anti-authority stand?

There is a noticeable absence of the word love in the discussion of sexual mores which I found limited. Puritanism and its hypocrisy is gone. But something has happened to excitement, shame and frustration. They don't "make love." They "make it." Brotherly love, shared drug deliria, but what about intimacy and sex?

Does Dr. West have any comments on how the "child-centered" culture of the past two decades appears in this rebellion?

Religious mystics have always "dropped-out" and "turned-on" and "tuned-in." They end up being truly more conservative, less innovative than the establishments from which they have withdrawn. Are the hippies becoming truly fundamentalist and simplistic?

In summary, the lucidity, the clarity, the benign skepticism of this chapter are useful and impressive. I hope it is a sign that we can expect more detailed information about individual true hippies in the not too distant future because of the evanescent character of this Green Rebellion.

THE DETROIT UPRISING: A PSYCHOSOCIAL STUDY

By CHARLOTTE DARROW, M.A. AND PAUL LOWINGER, M.D.

THE NEGRO in Detroit is considered to be better off than those in any of the large cities, and certainly he is ahead of the small towns and rural areas. This means that there is more individual home ownership, more college graduates, higher annual incomes and a good deal of participation in community life. For instance, there are more members of the NAACP in Detroit than in any other American city. During World War II there was a real incentive to come to Detroit in terms of the relatively high salaries paid in the automobile industry, which did not require special training and accepted the Negro. The union organized the Negro and gave him a responsible role.

The Detroit Negro community is characterized by not having a ghetto like the South Side of Chicago or Harlem. This is not to say that Detroit has open housing. There are many all black neighborhoods including the one in which the disturbance started. But Detroit has spread Negro home ownership about the city in a way that differs significantly from Chicago and New York. There are few neighborhoods in Detroit that do not have Negroes nearby or scattered about, which is not true of most northern cities. The successful, well educated, affluent Negro lives on the West Side which is where the riot occurred. Some middle-class Negroes on the West Side live near Twelfth Street in close proximity with poor Negroes of whom there are a great many. Twelfth Street, where the disturbance started has a reputation as a kind of Broadway: that is, prostitutes, drinking, pimps, gambling, narcotics and marijuana.

After hours drinking places—blind pigs—are often raided in Detroit, but it is usual to raid the black and not the white ones. Such a raid took place on Sunday, July 23, 1967, in the early morning, without a single precipitating incident of police brutality. Following this, the crowd began to gather in the street and a localized rebelliousness began.

Rumors went about that there had been a misdeed by the police. The crowds surged into grocery stores where they first took the expensive meat, champagne, and canned goods and then cleaned the places out. Liquor stores, pawn shops, furniture shops, and appliance stores were next. There was a carnival atmosphere on Twelfth Street. People spoke of "shopping for free." Furniture warehouses were looted in an assembly line fashion in which people passed the large appliances out and were courteous to each other. Some white groups joined in the looting. People who took more than they needed drove to Chicago, sold the loot and came back and got more from other stores. After a store had been looted, very often it was set on fire. There were many ways of setting stores on fire but the most common was the use of a tar and gas mixture which guaranteed that the fire would last long enough to burn down the store. Initially, the Detroit police were overwhelmed and were inactive; when called, the State Police, the National Guard and finally the federal army came into Detroit. The toll for the week was 43 dead, 33 black and 10 white; $100 million in property damage; 7200 arrests; 682 burned buildings and thousands of injuries. By Friday there was some semblance of normal business activity. The loss of production from automobile plants, industries and offices is incalculable as the city was shut down for a week. The city that had produced the union-management violence of the thirties as well as wartime race riots in 1863 and 1943 was in trouble again. The nonviolent amity of the 1963 Detroit Walk to Freedom by 125,000 blacks and whites was shattered.

This study was planned during the disturbance; the interviewing began on August 1 and was complete by August 20. The principal interest was in those most active in the uprising so we studied the committed young Negro activist. A structured interview explored what he did day by day during the disturbance, his thoughts and feelings about it, his opinions on Martin Luther King in contrast to H. Rap Brown, what about the war in Vietnam, what did the uprising mean, what should be done in Detroit now, would it happen again and other issues. The 47 questions (See Appendix) were mostly open ended, and the interviewers took down the whole response.

The 12 interviewers were young Negroes—men and women between the ages of 20 and 30—who were themselves active in the disturbance. They came from a black power organization, the staff of a new activist Negro newspaper and black militant student groups. Some of the interviewers and the project directors knew of each other through the Civil Rights Movement. The interviewers were paid but the subjects were not. We all met at a student's apartment in a neighborhood housing project where some of the interviewers live. In the days following it became the staff headquarters. We began training sessions with the interviewers. At first this meant establishing trust between the interviewers and the directors. We did interview training through the use of role playing; we interviewed each other. Trust of each other and of us was manifest in their personal answers to the questionnaire; our interviewers are in our sample. But no one knew if we would get true responses on the street. Each interviewer made five or ten pretest interviews and were convinced that if they were not getting the whole truth, what they were getting were true answers.[1]

Our sampling technique was to ask our collaborators to interview their male Negro friends and acquaintances between the age of 15 and 30. In order to cut down on the errors of response and nonresponse, and to address the problems of sampling a universe of the young Negro activist, we turned to the interviewer, the interviewing situation, some features of interviewer effect and to the sociological phenomena found in clustering.[2] When interviewers are less competent or are less experienced, group membership similarities might substantially assist the interviewer in maintaining rapport.[3,4] To get the quality of information we were all interested in, we assumed that our interviewers would seek out those people with whom they felt most comfortable: those most like themselves. Clustering refers to the fact that individuals associated with a given human group resemble one another and the tendency to homogeneity holds for most characteristics. Further, given a set of elements, the greater the segregation, the greater the homogeneity within the unit.[5] Finally, there was the problem of anonymity. As far as the directors were concerned, our respondents were completely anonymous. We did not ask for names or addresses, but only for the location—the cross streets nearest—of the home. This was because we were interested in the socioeconomic aspects of the neighborhoods. But while anonymity frees the respondent from fear of reprisal, it also seems to free him of inhibitions about inflating his prestige.[6] Considerable attention to the reliability of the reports of the activity of our subjects during the riot made us fairly secure about this. However, our interviewers felt that one of the places where inflation of prestige would most likely show itself would be in reporting the degree of education. Our interviewers were almost all college students or college educated, so it would seem that the respondent would try to make himself as much like them as possible. At the same time they would understand, and not-so-subtle signs of education would appear among those they interviewed.

Jobs and work are other areas of potential distortion and inflation. Some young Negro males prefer to give the impression to their friends and peers that they are not working when in fact they are. It turns out that *because* having a good job is so important, an opposite effect takes over and there is a bravado about not working. A Negro may say, "I don't have a job. I don't work at something that is degrading and humiliating. But still, I have good clothes and a good car." We asked our interviewers to comment on the motivation, interest, the hostilities, the apathies, of the people they talked to. We worked together for a period of three weeks during which the directors interviewed the interviewing staff in some depth.[7]

Our category of activism includes followers of Stokely Carmichael, H. Rap Brown, Malcolm X, Floyd McKissick and also some who follow Martin Luther King. Of course, there are local Detroit Negro leaders and groups such as the West Central Organization and the Inner City Student Committee which are activist, in contrast to the more established older nonviolent NAACP approach to community problems.

Most of the 222 people we interviewed followed a pattern. They were unmarried, urban, Negro men. There were a good proportion of students and intellectuals. Most of them were unmarried, and the married had few children. The data fall in three age groups: 15 to 19, 74 subjects; 20 to 25, 98; 26 to 30, 50.

The teen-agers were generally in the right grade for their age; 33 per cent were entering their senior year in high school; 53 per cent were in college. This contrasts to the study of 670 teen-agers (13-17 years old) arrested during the riot who were one to two years behind in school grade for their age.[8] Among the 20-25 year olds, 70 per cent had finished high school, including 14 per cent who have completed college. Over 25, there were 72 per cent who had finished high school including 20 per cent college graduates. This contrasts with 25 per cent Negro high school graduates including 7.4 per cent college graduates in 1966.[9] Among the 20-25 year olds, 65 per cent were single and of the 26-30 group, 42 per cent were single. It should be noted that in our sample there was considerable unemployment among the two thirds who were nonstudents: 70 per cent between 15 and 19; 40 per cent between

20 and 25; 24 per cent from 26 to 30. It is difficult to come to any conclusions about this because July 23-28 was in a week when automobile models were being changed from one year to another and there were many layoffs. People with low seniority, which includes many Negroes, would be likely to consider themselves unemployed since they could not be sure they would be called back to work when the model change over was completed.

Most of the sample, 192, (86 per cent), were on the streets during the disturbance. A third of our subjects (76) were participants who looted, used weapons or set fires as distinguished from the two thirds who swelled the crowd as spectators or stayed home. There were 30 subjects (14 per cent) who were neither participants nor on the street but who stayed home except for going to work or other necessary travel. A third (26) of the 74 teen-agers took part in the looting and three, 4 per cent, also set fires. Among the 98 men in their early twenties, 40 per cent (39) took part in the looting. Also there were five arsonists and five snipers, that is, 10 per cent in this group. The 50 oldest men, 26 to 30, were the most cautious. Half of this group was on the street but only nine members, 18 per cent, took part in the looting. It was this age group who were most upset by the events. Negative feelings about the riot were reported by 58 per cent, while only 18 per cent had positive, excited feelings with 24 per cent who were ambivalent or undecided. In contrast, among the teen-agers, 50 per cent had positive feelings: happiness about being able to loot and feelings that at last the Negro was fighting back and getting something from "Whitey." Negative feelings included fear of the police and anger that they couldn't get as much loot as they wanted. In the 20-25 year old age group we encountered a great many deep personal and political feelings. Positive feelings amounted to 52 per cent among the 20-25 year olds while only 16 per cent negative and 32 per cent were ambivalent. Most of the positive feelings were about the success of the rebellion, the assertion of rights by the Negro and burning of stores of bad business men. "Bad merchants," usually white, were those who had not extended or had abused credit, those who sold inferior products for high prices and held the community in contempt. The hope was often mentioned that in the burned stores, Negroes would replace the white owners. The snipers were invariably shooting at the enemy, the police, and said they knew exactly what they were doing.

The subjects were asked whether they agreed more with Martin

Luther King or H. Rap Brown. The viewpoint of the people who could choose was largely that of H. Rap Brown by 81 to 62. The men in their early 20's were 64 per cent for Brown, the teen-agers were 56 per cent for Brown but only 40 per cent of the oldest men agreed with Brown. A third of the entire group of 222 were undecided, had some agreement with both or made a different choice. These Negroes are no longer interested in "marching around in the streets for nothing" as one man put it. In this connection, it is of interest that 20 per cent of the sample had participated in the Civil Rights Movement.

Those with an opinion about the war in Vietnam held that it had an effect on the unrest in the Negro community in 45 per cent of the answers, while 55 per cent did not hold this view. This is probably an underestimate because the question on whether the war in Vietnam had an effect on the riot was sometimes misunderstood. Some people replied that the war had no effect on the riot but said in response to a question on how the war affected them that the black man had no business in Vietnam.

Among the 153 subjects who expressed an opinion about whether there would be another riot, 90 per cent said it would happen again although they were unable to predict when it would happen.

A 22 year old office worker who is a high school graduate describes his experience on Thursday as follows: "Some guys were breaking into a clothing store across the street from the Algiers Motel and the police came and started shooting before they cried 'Halt'. The gun clicked twice and two Negroes were laying on the ground after the shooting ended. I got scared and ran. The police saw myself and a friend running down the alley. They called 'Halt' but I didn't stop. So the police started shooting. I was hit in the back and was in the hospital for four days." A 22 year old married factory worker who is an active Catholic asked, "How was your home affected?" He says, "My apartment was shot into by Guardsmen or police." Did you set fires? "Yes, also I burned a store on Mack. Hated the bastard that owned it for calling the cops on me once." Did you see shooting? "Yes. When the cops fired into the apartment where my family was, I got my rifle up on the roof and I'm sure I hit at least one or two." A 21 year old married factory worker who finished 11 grades: What did you do on Wednesday? "I went to work on Wednesday after the curfew was over, but on my way I was stopped by the police, told him where I was going but he still made me get out of my car, searched it, searched me, checked my license. While all this was going on, I was called 'black nigger, mother-fucker' and every name under the sun. Finally, they let me go. But before they did, they said "We're going to kill all the smart niggers." A 24 year old student: How did the disturbance make you feel? "Extremely happy." Who do you think the riot was against? "White power structure; the evils of society." What do you want done now in Detroit? "Neighborhoods wiped out should be rebuilt and controlled by Negroes." A 22 year old Negro student who drives a truck: Who do you think the revolt was against? "Basically directed against white property and white state power and in a broader sense, some black property owners too; directed against the entire concept of the American way of life." An 18 year old high school junior who is unemployed is asked what do you want now in Detroit? "The white man let me work, and get his foot off my back." Did you see looting? "Yes, I did quite a bit, a pawnshop, supermarket and a warehouse too. Also, I saw a policeman looting a liquor store on Jefferson." Did you see fires, firesetting? "Yes, but I didn't set any, although I was offered money by a white man to burn his store after it had been looted." Who do you think looted, shot or set fires? "Police and poor people and hippies." A 25 year old Catholic college sophomore answers the question: What do you want done now in Detroit? "That everyone be treated right." A 16 year old ninth grader who is employed in a summer job answers the question: Have you ever been in the Civil Rights Movement? "No, and don't want to be. They don't accomplish anything."

Certain conclusions are apparent from the interviews. The lack of communication between our subjects and the power structure is clear. This is no simple problem to be solved by calling a biracial committee. This new wave is not communicating with the Negro or the white leadership, and the lack of communication is apparent in several areas. The Civil Rights Movement as we have known it is largely dead or moribund. No one denies the historical importance of Martin Luther King who now leads a significant segment of the American peace movement, nor questions the real achievements of the civil rights activities, especially in the South. But the local and national leaders of the Civil Rights Movement no longer speak for the militant segment of the Negro community we have interviewed. The new leadership is largely informal and almost entirely unrecognized in either the white or Negro establishment.

The finding that the events in Detroit were neither a riot nor revolution is worth mentioning. A riot is usually defined as aimless behavior involving disturbance or tumult. A revolution is the overthrow of a government by a new political leadership with a new program. This lies somewhere in between. To call this a riot is to miss its significance and to call it a revolution is to overestimate its structure and directedness. We characterize it as an uprising since it represents a spontaneous insurgency of slum dwellers who feel oppressed by the political and economic structure. They are comparable to the crowd in the French Revolution which was drawn from the petit bourgeoise.[10] They are not Moynihan's "underclass" although some black lumpenproletariat are present. It is to be noted that these slum dwellers include a significant proportion of intellectuals, students, and professionals who are militant activists and opposed to the power structure. This is crucial since the activists have developed an ideology which infuses nearly all the participants regardless of background. It will be interesting to see in further studies whether the same ideology which pervades 15 to 30 year old men also affects the female and the older male. Perhaps some of the white participants in the riot may show a similar attitude toward society based on a different set of experiences and grievances.

It has been suggested that the disturbance in Detroit was against private property or capitalism. There was little in the replies of our subjects to suggest that they wanted socialism, communism, or egalitarianism. They want a better distribution of the goods of our society, but voiced little basic quarrel with the economic system.

Some have said that the disturbance was biracial and did not include hostility between whites and Negroes. Our study indicates that this Negro struggle is largely one of racial self consciousness, and it was almost accidental that there was not more hostility between black and white. The looting by whites including college students is under study and will be reported.

The ideological basis of the uprising remains preeminent and is the least recognized aspect of the disturbance. The main demand of the people who answered our questionnaire is for self government. Self government means control of the neighborhood in which they live. Self determination represents something that is out of keeping with the American political and economic system as we know it. It is the limitation of our constitutional and economic imagination that does not permit us readily to perceive what is being asked. There is a demand for black businesses in the ghetto: those owned and operated by Negroes.

Until now these businesses have been 90 per cent white-owned with some Negro employees. How can this come about by the destruction of present stores? A militant New York Negro leader explained it this way, "We burned the stores in Harlem in 1964 and the Jews didn't come back and now black people own them." Whether this will happen in Detroit should be the subject of a study, but it is a part of the local ideology. It is clear that our subjects do not regard the presence of two Negro congressmen, a Negro City Councilman, Negro state legislators, black judges and other Negro officials in Detroit as control. Perhaps real control means electing the police sergeant who runs the station or having a community committee select him. Perhaps control means a board of citizens who will select a Negro school principal. Certainly, if Negro neighborhoods have Negro businesses, there will be more jobs for Negroes. How Negro employment is to be fostered in the industries which stretch across state as well as city and neighborhood boundaries is not clear from our questionnaires but certainly is an issue. The assertion that an ideologic demand for autonomy is the primary consideration of the disturbance, is not to deny the importance of good schools, good housing, more jobs, less unemployment and effective and tolerant police. Our subjects named these issues when asked what they wanted done now in Detroit. The almost exclusive emphasis on these considerations offered by government officials from the President on down is the hope that money can buy pacification of the ghetto. Our study would indicate this is a futile hope despite the fact that such a program must be a part of the change in American cities.

What about integration, the dream of black and white together, so important to the 1954 Supreme Court decision on education and the years of civil rights activity? It has become irrelevant according to our subjects, who largely ignore it. Integration simply does not appear among the demands for power, control, and autonomy nor in the methods suggested to attain these goals. Better schools, jobs, police, and housing which are mentioned are not necessarily integrated schools, jobs, police, and housing to our subjects. Does this mean that an all black program is the goal? Integration versus segregation is now being actively debated in the inner circles of the new black power.

Another aspect of the ideological struggle is the creation of a new self image for the American Negro. The need to dramatize self assertion and independence in one's own life is no less important to the subjects than to control their own communities through collective action. Dr.

Poussaint, a Negro psychiatrist, writing in the New York Times describes one of the main problems of the American Negro as self hatred, an aggression-rage constellation which turns inward and becomes the well known Negro docility, passivity, and nonviolence.[11] The feelings in the Detroit uprising may be similar to those reported by Frantz Fanon, another Negro psychiatrist.[12] Recall that Fanon suffered from racism and the feelings of racial inferiority without ever setting foot in the U.S., since he was born in Martinique and lived in France and Algeria. Fanon's assertions are the same as those of our subjects: In order to overcome the colonial mentality, the feeling of being "a nigger," black people must express and act out feelings of anger and violence. If Fanon is correct in his assertion that the struggle of a colonial people for freedom must be accompanied by violence, then his prophecy may apply to the Negro in the American city. Like the blacks of Fanon, some of our subjects assert their manhood through looting, shooting and setting fires and find it a liberating experience.

The destruction of property and lives and the violation of law involved in an uprising may be seen as symbolic assertions of manhood which cannot be met alone with reparations such as the billion dollar A. Philip Randolph plan for the cities or the Hubert Humphrey-Marshall plan for the Negro. A facetious solution is the suggestion that there should be a yearly Riot Day for which merchants could budget, and this might insure a better attention to community relations than the present biracial committees.

One of the most important assertions of our study is that this type of disturbance will happen again in Detroit. The present uprising did not solve the problems but indicated the power of the participants. The informal leadership who use the term black power now feel their strength. Next time, we are told the disturbance will be different. The new disturbance will aim at vital public services such as electricity, telephone, gas, radio, water supply, etc. This would be seen as a further assertion of manhood and power and a new demand for self government. If the needs are met in some fashion, it is possible the second phase of the disturbance will not take place despite the fact that many of our subjects fear it, yet hope for it and expect it.

What is the role of psychiatry and the social sciences in a situation of this kind? First, we might reply hastily that there is no role at all other than as interested, concerned citizens. Some of the black leaders who have talked with us asserted this in response to our initial request for their assistance. On second thought, perhaps there is a role. The social sciences can take the views of the Negro revolutionary and their followers to the white community. The Negro revolutionary does not appear readily before investigating groups of either the academy or government. The ability to bridge this gap must be in part, the responsibility of social scientists. This role will not be met by the Civil Rights Movement, the Negro middle class, the government, or the press. Psychiatry, in particular, may understand these matters because its daily work is to translate the irrational into the rational and understandable. To many Americans an uprising of black people in an affluent, democratic Christian nation seems as incomprehensible as the nervous cough of Freud's hysterical patient Dora did in 1905. Freud thought the cough understandable as a substitute for a repressed desire by Dora to suck her father's penis. So the occult seems reasonable despite the "astonishment and horror" which Freud says his explanations stir in the reader.[13] Psychiatrist Coles says it this way: ". . . the law which rioting Negroes now defy once decreed them to be chattel . . . Negroes . . . are a nation within a nation, and until that awful division of territory, money and power is somehow bridged . . . rioting will continue."[14]

An alarum seems more appropriate than a summary. A 29 year old professional man who was a sniper and a looter and set fires said the Detroit uprising was against The Great Society. Two weeks later Senator Fulbright wrote, "The Great Society is a Sick Society."[15] Mayor Cavanaugh of Detroit said, "What will it profit this country if we, say, put our man on the moon by 1970 and at the same time you can't walk down Woodward Avenue in this city without some fear of violence?" The same concerns led the Governor of Michigan to this imperative: "I am more convinced than ever before that unless we reverse our course, build a New America, the Old America will be destroyed."

REFERENCES

1. Katz, D.: Do Interviewers Bias Polls? Public Opinion Quarterly, 6:248-268, 1942.
2. Rose, A.: Public Opinion Research Techniques Suggested by Sociological Theory. Public Opinion Quarterly, 14:204-214, 1950.
3. Feldman, J., Hyman, H., and Hart, C.: Interviewer Effects on the Quality of Survey Data. Public Opinion Quarterly, 25:749-750, 1951.
4. Stember, C. and Hyman, H.: Interviewer Effects in the Classification of Responses. Public Opinion Quarterly, 13:680-682, 1949.

5. Kish, L.: The Selection of the Sample. *In* Festinger, L. and Katz, D. (Eds.): Research Methods in the Behavioral Sciences. New York, Holt, Rinehart and Winston, Inc. 1965, pp. 210-11.
6. Hyman, H., Cobb, W., Feldman, J., and Hart, C., and Stemper, C.: Interviewing in Social Research. Chicago, University of Chicago Press, 1954, pp. 182-184.
7. Hyman, H.: Survey Design and Analysis. Glencoe, Free Press, 1955, p. 162.
8. Komisarut, R. and Pearson, C.: Juvenile participation in the Detroit riots. J. Urban Law. *In press.*
9. U.S. Department of Health Education and Welfare: Social and Economic Conditions of the Negro in the U.S. Oct., 1967, pp. 46-47.
10. Rudé, G.: The Crowd in the French Revolution. New York, Oxford University Press, 1959.
11. Poussaint, A.: A Negro Psychiatrist Explains the Negro Psyche. The New York Times Magazine, Aug. 20, 1967.
12. Fanon, F.: The Wretched of the Earth. New York, Grove Press, 1963, pp. 27-74.
13. Freud, S.: Collected Papers, Vol. III. New York, Basic Books, 1959, p. 59.
14. Coles, R.: Maybe God Will Come and Clean Up This Mess. The Atlantic, 220:103-106, 1967.
15. Fulbright, J.: The Great Society is a Sick Society. The New York Times Magazine, August 20, 1967.

APPENDIX*

This study is being conducted by the Lafayette Clinic and Department of Psychiatry of Wayne State University.

NAME OF INTERVIEWER:

DATE OF INTERVIEW:

NEIGHBORHOOD OF RESIDENCE: (use main street intersection)

WHERE AND WHEN INTERVIEWED:

DESCRIPTION OF SUBJECT:
AGE: SEX: RACE: MARITAL STATUS: SINGLE ________
MARRIED ________ DIVORCED, WIDOWED OR SEP. ________
NO. OF CHILDREN:
EDUCATION OF SUBJECT: (no. of grades) ________
OCCUPATION OF SUBJECT:
UNEMPLOYED OR LAID OFF THE WEEK BEFORE: YES ________ NO ________
RELIGION: (C, P, J) ________ ACTIVE OR INACTIVE: ________

DESCRIPTION OF BEHAVIOR:
WHEN AND HOW DID YOU FIRST HEAR ABOUT IT?

WHERE WERE YOU?
HOW DO YOU THINK IT STARTED IN YOUR AREA?

WHAT DID YOU DO? (include details)

SUN:

MON:

TUES:

WED:

*To conserve space in this publication, the spaces for answers in original form have been condensed.

WHAT DID YOU DO?

THURS:

FRI:

WHO DID YOU SPEND TIME WITH?
AGE: ________ SEX: ________ RACE: ________ FAMILY OR NOT: ________

HOW WAS YOUR HOME AFFECTED?
DID YOU SEE LOOTING? (give details)
DID YOU SEE FIRES? FIRESETTING? (give details)
DID YOU SEE SHOOTING? (give details)
WHO DO YOU THINK LOOTED, SHOT OR SET FIRES? (give details)
WAS THERE ANY DIFFERENCE IF A BUSINESS WAS WHITE OR NEGRO OWNED?
WHAT KIND OF BUSINESSES WERE MOST OFTEN DESTROYED?
HOW DID IT MAKE YOU FEEL? SAD, FRIGHTENED, EXCITED, HAPPY, ETC.?
WHAT DID YOU THINK ABOUT?
WHAT OTHER EXPERIENCE WAS LIKE THIS?
DID YOU AGREE MORE WITH MARTIN LUTHER KING'S STATEMENT OR WITH THE VIEWS OF H. RAP BROWN?
DID THE WAR IN VIETNAM HAVE ANY EFFECT UPON THE RIOT?
HAS THE WAR HAD ANY EFFECT ON YOUR LIFE?
ARE YOU A VETERAN?
WHO DO YOU THINK THE RIOT OR REVOLT WAS AGAINST?
DO YOU THINK IT WOULD HAPPEN AGAIN?
WHY? WHEN? WHERE?
WHAT DO YOU WANT DONE NOW IN DETROIT?
HOW DID THE POLICE ACT? (give details)

HAVE YOU EVER BEEN IN THE CIVIL RIGHTS MOVEMENT?
IF SO, WHAT DID YOU DO?

Discussion by Edwin A. Weinstein, M.D.

What has brought the attention of the world to the Detroit uprising and similar disturbances is the violence. Most people in this country approve of the Negro's ambition to have better jobs (provided it is not their job) and more education, but not of the method of attainment. The objection is not to violence per se because as long as Negro mayhem is restricted to other Negroes, few people become upset. What is now happening is that the violence is reaching white society. I recently heard a talk by the Black Power leader, my ex-colleague at Howard University, Nathan Hare, in which he speculated on what revolutionary results could be obtained if black women started cutting up white men instead of black men.

The authors note that the uprising did not occur in a black ghetto and that among the participants were people with fairly good jobs and considerable education. Also, there were people in the area who did not participate. These facts are important because violent protest is not the inevitable sequel of social and economic deprivation, even though the protester may feel it as such. Here the authors may be taking a too simple view of cause and effect. Then, poverty is a relative matter, despite the arbitrary figure that the government has set. It is relative to what others have and to one's aspirations. Revolutions generally are not made by the lowest classes even though they may provide the action.

Psychoanalysis is concerned with meanings and motives and relationships among people. Human beings rarely define events, within or without the body, in terms of their intrinsic physiological or physical properties. Rather, events take on form or meaning by reason of their anticipated consequences and in terms of a socially organized pattern. For example, Abraham Wikler, the psychiatrist and pharmacologist, has commented that LSD was given experimentally for years to subjects at the U.S.P.H.S. Hospital in Lexington without anyone ever having taken a "trip." Rather, the experience was described in terms of disordered perception and generally unpleasant bodily sensations. It is only when the taking of LSD is part of a group ritual in a society that has a culture and a language that people take "trips." We uncommonly perceive the meaning of violence as violence but rather in terms of purpose and group identification. Thus, the Detroit informants variously perceived their actions as "exciting," "against Whitey," as revenge on brutal cops and dishonest businessmen, and as establishing a black community (even though it involved the physical destruction of the neighborhood).

It may be helpful to classify the violent behavior of the Detroit participants in terms of language or modes of communication—bearing in mind the meaning of communication as sharing. Elsewhere in this volume, Spiegel notes the unity that appears after violent action. The first category can be called rioting. This is the participation in a euphoric, emotionally stimulating experience and the enjoying of a "riotous" good time. It imparts an ecstatic sense of identification with the Negro group. The second mode of communication involves a paranoid identification (here I am not implying psychotic behavior, since paranoid delusions and formulations are generally true if one treats them metaphorically) with whites with feelings of hatred, envy, revenge, etc. A third kind of communication is referential rather than experiential and is the use of explicit threats and acts to send the messages that we are now getting from the Black Power advocates. This classification is hardly mutually exclusive as people may communicate in a number of ways at the same time.

I recently read Eric Lincoln's excellent book on the Black Muslims. The Black Muslims, who, interestingly, developed in Detroit, are seen as an exotic anti-white, anti-integrationist and anti-American sect. Yet, many of the teachings and practices of the Muslims could come out of Benjamin Franklin and Horatio Alger. These include working steadily, saving money and accumulating property, wearing conservative clothes, keeping slim, refraining from drugs, preserving the virtue of women and establishing the authority of the man in the home. The Muslims advocate obeying the law and using violence only in self defense. Even its opposition to the Vietnamese war in the person of Muhammed Ali has been on the grounds that he is a minister of religion. It is essentially a conservative group that wants the black man to return to the pristine purity that he had before the white man corrupted him.

The authors talk about the creation of a new self image for the American Negro. This concept needs a great deal of clarification. If it is an ideal, it seems to me rather pathetic to limit one's self to the physical circumstances into which one is born such as race, religion and gender. The great accomplishment of the Peace Corps is that it has enhanced in its members an identity that has transcended skin color, sex, religion, social class and even, in a sense, nationality. It is, though, unreasonable to ask the Negro to do what members of other ethnic groups, Polish-Americans, Jewish-Americans, etc., have not done.

Another point relative to Negro identity is the system of classification of race. Here there is a contrast between that used in the United States and in other parts of the world, such as the West Indies. In America, there is an arbitrary division into white and Negro, regardless of actual skin color. In the Caribbean, skin color tends to be descriptive rather than classificatory, so that people are perceived as black, brown, red, pink, etc. Until relatively recently, the term Negro was used infrequently.

We are in the midst of a revolution and I agree fully with the authors that what we call the Negro problem is the problem of us all. It is no more irrational for the Negro to seek his way out of his difficulties through violence than it is for the country to seek peace by killing Vietnamese. The Civil Rights Movement may be dead, but it has succeeded to a great degree. It has not, however, solved economic problems or made the Negro immune to the frustrations of our society as a whole. Much of the violence directed toward the upper classes of whites and Negroes is an expression of rage and disappointment that they don't have the answers.

Part IV. Social Dynamics

THE SOCIAL AND PSYCHOLOGICAL DYNAMICS OF MILITANT NEGRO ACTIVISM: A PRELIMINARY REPORT

By JOHN P. SPIEGEL, M.D.*

INTRODUCTION

STUDIES of the relation betwen personality structure and political leadership have been carried on for many years but in a somewhat limited fashion. The research area is a subdivision of the more general topic concerned with the relation between personality types and occupational choice. Professional political activity, however, is a more sensitive topic than most other occupations. The difficulty in gathering systematic data in so touchy a subject matter perhaps accounts for the fact that research in the area has not been pursued more vigorously.

No attempt will be made here to review the research literature on this topic. Suffice it to say that contributions have been made from a broad spectrum of disciplines: for example, from political science,[1,2] economics,[3] social psychology,[4] sociology,[5] as well as from a psychoanalytic orientation.[6,7] The present report on some personality mechanisms involved in the motivations of Negro militant leadership is based on psychoanalytic considerations. However, its limitations are greater than usual, and they should be made clear at the outset.

During the year that the Lemberg Center for the Study of Violence has been in existence, its staff has been studying the urban process associated with outbreaks of violence and rioting in a number of

*Director, Lemberg Center for the Study of Violence, Brandeis University, and Professor of Social Psychiatry, Florence Heller School for Advanced Studies in Social Work, Brandeis University, Waltham, Massachusetts.

northern cities. We regard the phenomenon as a collective response within the ghetto community to frustration and conflict resulting from the social behavior of whites toward black people. Our main research concern has been to determine what factors in the behavior of white power structures and in the responsive orientation of black communities are correlated with major rioting. Why, we have asked, do some cities experience so much more retaliative violence (or revolt) than others? Are there significant differences in the way ghetto residents perceive the mayor, the police, the political parties, and the business community, and the welfare establishment in a riot city versus a city that has not had a riot?

In the course of gathering data on these questions in six northern cities, our staff has interviewed forty to fifty black and white "decision makers" in each city. A good many of the Negroes interviewed were ghetto leaders of various degrees of militancy. They were asked to give their views on how race relations were being handled in their city, on the possibility of a riot if none had occurred, and on the causes and ways of preventing riots if one had occurred. In addition to this systematic interviewing, the staff, which consists of both black and white professionals and nonprofessionals, has been in informal contact with black nationalist and black power leaders in seven other cities.

The limitation in what I have to report in this paper is that these interviews were not designed to elicit personality variables. Even so, motivations and personal factors, as one might expect, appeared prominently if unsystematically, in the course of the interviewing. Since the data were not systematically gathered and the methods used were not technically geared to elicit psychodynamic processes, the interpretation of the interview data is impressionistic and must be treated with caution. The main reason for reporting our impressions at this time is that so little is known about the new crop of black, militant ghetto leaders. Reports from the press and the electronic media have tended to produce a rather undifferentiated, stereotyped image of these leaders.

There is one further reason for making a preliminary report now. Styles of militant leadership in black communities are currently in a state of considerable flux. A year or two hence, leadership roles and the types of personalities recruited into these roles will undoubtedly have undergone change. They may even have crystalized into definitive patterns by then. It this should turn out to be the case, a report on the developing and changing situation now, no matter how incomplete, can make a contribution to our knowledge of the relation between personality structure and political leadership based on the interaction of each side of the equation in a changing situation. In other words, it may be easier now to determine whether the direction of change is influenced more by personality factors than by environmental factors—or vice versa—than if one waited until after crystalization and fixation of leadership roles have taken place.

The Phenomenology of Negro Militant Leadership

The current situation of flux is characterized by the presence of several partly competing, partly cooperating styles of leadership in black communities. At the present time it is impossible to foresee which of these styles will win out over the others and set the pattern for the future. Nor can one know whether new leadership styles not yet in evidence may turn out to be of greater significance than forms of political behavior currently visible. Innovation is the name of the game now being enacted in the ghettos.

Before distinguishing the several styles of militant leadership, we should be clear about what they all share in common. The term, "Black Power," has been generally accepted as designating the core philosophy of the new militancy, but it has been too loosely used—it is semantically too diffuse—to serve without qualification as a useful description of the orientation of this leadership. What the militant leaders share is: (1) a disillusionment with the possibility of obtaining adequate representation for black people in the economic processes of the overall white society; (2) a concern for the weakness of organization in black communities and a desire to mobilize group consciousness and a feeling of identity for the sake of self help, that is, for solutions of their problems instituted by black people themselves; (3) a desire to create a political program and a style of leadership capable of realizing the aims of self help and community development.

Coordinate with these aims is the widely accepted opinion among militant leaders that such leadership entails severe risks. As Stokely Carmichael and Charles Hamilton phrase it in their book on *Black Power*: "Those who would assume the responsibility of representing black people in this country must be able to throw off the notion that they can effectively do so and still maintain a maximum amount of security. Jobs will have to be sacrificed, positions of prestige and status given up, favors forfeited. It may well be—and we think it is—that leadership and security are basically incompatible."[8] In our interviews, respondents went beyond the economic and occupational dangers to describe the possibility of physical injury or death from hostile forces within both the white and black communities. Some of them displayed an attitude symbolized in the gladiator's cry: "We who are about to die salute you!" (*Morituri te salutamus*!) For example, they told interviewers to get as much information as possible as quickly as possible since they might not be around much longer to be interviewed.

Finally, as a result of their disillusionment and growing alienation from white society, the new militant leaders share a refusal to compromise their aims at the request of liberals or others in the white establishment speaking in the name of "sweet reason" or political practicalities. In his book on the recent history of the black protest movement, Fred Powledge has described the dangers to the realization of the aims of the movement which occur when Negro leaders accept positions in white organizations, allowing themselves to be "co-opted" or to believe in the possibility of making progress through peacemeal measures.[9] Resistance to "tokenism" and all other "take it easy" or "go slow" policies combined with offers of white help are basic elements of the militant position.

Within this configuration of shared attitudes, militant leaders differ in their political styles and their programs for achieving the aims of the protest movement. Variations occur in the degree to which black leaders remain invisible and underground or are plainly heard and seen within the black community (to say nothing of their accessibility to the white community!). Differences occur in style of dress and speech, in religious orientation, in the use made of other nations and peoples, such as Moslem, African, or Far Eastern (Asian) countries for purposes of identification, and in the willingness to use violent technics to ward off the "legitimate violence" of white communities and their law-enforcement agents employed as forces resistant to change. Finally, there is variation in the social class and occupational backgrounds of the militant leaders. We shall use this last variation, plus degrees of visibility, as the major axes of differentiation, distinguishing four categories of working-class and a group of middle-class militant leaders. It must be understood, however, that the distinctions are abstract and are apt to break down in actuality due to the amount of overlap between categories.

First, as to the leaders from working-class or lower-class backgrounds, the four types are: The Black Nationalists, Community Action Workers, Street Leaders, and Revolutionaries.

The Black Nationalists

These are men whose primary orientation is ideological. They spend a great deal of their time formulating histories of the black people, paying particular attention to neglected cultural matters. They believe that a sense of pride and a constructive feeling of identity among black folk is dependent upon their being correctly taught their own history, free of the distortions and anamolies imposed on their people in white schools. Thus they are essentially teachers. Some of them emanate a sense of mystical identity in their manner of teaching. The more mystical black nationalists describe mythologies of vague origin and traditions which cannot be easily documented. They vary in the sources of the traditions, some tracing them to Africa, others to the Near East or even to India and the Orient. They also tend to be self taught, having come upon their insights through reading and deep contemplation. This imparts an air of freshness, conviction, and discovery to their teachings.

Because of their working-class origins and the fact that many (though by no means all) have not completed their formal education, they usually present themselves as a man-of-the-people. They are the voice of the ghetto in the role of the Guru, the honored, almost saintly teacher. Because of their "guru roles," they are quite visible within the ghetto and their visibility is enhanced by their adoption of exotic dress: a pill-box hat, a bearded chin, a shaved head or "natural" hair style, long or short tunics open at the neck with an African "tiki" or some other ornament hanging from the neck. The dress varies among the different black nationalist leaders but their followers tend to adopt the style of the leader. The style of dress symbolizes both the rejection of white, middle-class costumes (and identifications) and the importance of an identity in line with the historical origins of black people. This identity is further underscored by the leader's adoption of a new name usually based on African or Arabian patterns. As with dress, the followers adopt a new name in the style of the leader.

Left to their own devices, most black nationalist leaders would teach history, African languages, and cultural studies in a "freedom school," or in an "academy" of their own. But they feel that the oppression under which black people live makes it impossible to give their full time to such activities. The people need to be led out of oppression through the organization of independent black communities or "nations," segregated from the overall white community. The leaders have as yet few economic or political blueprints for these communities, tending to leave such matters to others. Rather, they concentrate on preparing the psychological and social underpinnings for the future development of the black communities. Accordingly, although some of their thinking diffuses to the adults in the ghetto, they make a direct approach to the youth, to the Negro teen-agers who are so little under the control of any part of adult society. It is the young who need to be organized and given discipline, pride, and a sense of future direction. To facilitate the inculcation of discipline, followers are organized within bureaucratic structures which combine military titles with propaganda tactics, somewhat on the model of an election campaign staff. Followers are called captains and lieutenants, and function on descending lines of authority.

Of all the new, militant leaders, the black nationalists are the most vigilant in their dedication to Negro separatism for its own sake. Since whites are not expected to accept such a development peaceably, they believe that a violent confrontation with the white community is inevitable and that they must be prepared. But they are not inclined to discuss their plans openly—at least not with our staff of interviewers. Therefore we have no information on this topic. At the interpersonal level, however, they display the attitude of acceptance of risks and dangers mentioned above. In so far as we were able to observe, they are authoritative, didactic, and supportive toward their followers. Toward whites who are able to listen they are distant and intensely critical of white behavior patterns, but willing to explain their aims and purposes. Whites are, without exception, placed in the position of adversaries. If a white person expresses a desire to be helpful to the black community, he is told to use his energies to change the white community and to leave black people alone.

We have described the characteristics of the black nationalist leaders at some length because they represent the most clearly defined and idiosyncratic group of militant leaders. The remaining categories of leaders will now be discussed in briefer fashion.

Community Action Workers

In the ghettos of northern cities there exist a variety of "self-help" organizations. Although they may receive some assistance from public funds, they are primarily organized from within the ghetto to solve a variety of problems—in welfare work, job retraining, legal assistance, school bussing, and the development of small business enterprises. Though the heads of organization usually come from middle-class backgrounds, the staff of such organizations are young men and women from working-class origins. Before becoming interested in community action work they may have participated in the semidelinqent, street activities characteristic of the ghetto, and may well have acquired a police record. In addition, they have frequently worked in some part of the formal poverty program within the city.

The social histories of these young people have thrown them into contact with the ordinary, somewhat jungle-like, "culture of violence" in the ghetto on the one hand, and with the achievement oriented, middle-class cultures of bureaucratic organizations on the other. They have learned something about how white mechanisms of social control work. They are both impressed by and offended by these styles of control—impressed because of their success, offended because they usually work against black people. Accordingly, they are determined to adapt white, middle-class, bureaucratic, and administrative styles to the aims of black self development. They give much thought to plans for economic development within the community. They are quite ambivalent about participating in formal political activity as they have little hope of breaking through the barriers of the party system. On occasion, they may work for a Negro candidate for office, but, on the whole, they expect little help from participating in organizations in the white community. Their main efforts are directed toward mobilizing angry, uncommitted, or apathetic residents of the ghetto to work for community development. In this way they hope to forestall the individual and group violence that is likely to appear among disorganized, aggrieved, and despairing people.

Though not as conspicuous as the black nationalist leaders, the young community action workers are visible and known within their communities. They prefer not to wear middle-class dress but ordinarily confine their distaste to dispensing with jackets and ties. While they sympathize with some of the aims of black nationalists, they regard the leaders as ideologically too extreme and too unrealistic in their methods of opposition to whites. In face-to-face contact with white people, they are less secretive, more willing to argue and debate. In the presence of black nationalists, well-intentioned white people feel a sense of menace to their entire way of life and existence. With community action leaders, whites experience pointed opposition and searching criticism but are likely to conclude that an eventual accommodation is possible.

Militant Street Leaders

Most black ghettos contain gangs of teen-agers who live in the streets and engage in various kinds of aggressive activities. In some cities, these groups are small and informal; in others they have grown to massive proportions by incorporation of the smaller gangs. The large gangs are highly organized. In recent years the gangs, small and large, have to some extent given up street fighting and the usual semidelinquent activities in favor of somewhat more constructive pursuits geared to community organization and recreation. As their goals have been modified, the gangs have retained the affiliation of older youths, some of them married, who would formerly have passed out of gang membership. As a consequence of these changes, styles of gang leadership have also changed. The leader is no longer the strongest, toughest, most aggressive and most fearless boy. He is likely to have been selected because he is older, more experienced, and more ingenious in his dealings with the overall community than his competitors. Brawn, cunning, and skill in weaponry are currently of less importance to leadership than organizational talent.

Despite the modification of gang structure and goals, gang members are still apt to be seen by the police as hoodlums who should be chased off the streets. The fact that the leaders are interested in Negro affairs and protest technics is of no comfort to the police who see this development as combining criminal interests with black power agitation. As a consequence, it is wise for gang leaders to stay out of sight in order to escape police harassment.

But their invisibility is inversely correlated with their significance as militant leaders. The youthful gang members are among the angriest black people in the ghetto. They resemble Banquo's murderer who, when Macbeth questioned his resolution, answered, "I am one whom the vile blows and buffets of the world have so incensed that I am reckless of what I do." Their wrath is directed toward the world of adults and the white system of authority in a rather undifferentiated fashion, though the greater part of it is aimed at the police.

One function of the gang leader is to contain these hostile energies and try to direct them into reasonably productive channels. But, if a riot should break out in the city, it is the teen-agers who are the most active participants in the early stages, and the street leaders have to go along with the action.* After the rioting dies down, the street leaders often emerge in public for the first time to confront civic authorities and articulate the angry complaints of the teen-agers. Then the city authorities, surprised and caught off guard, wonder where these leaders kept themselves before "surfacing" and why no one had talked with them before the trouble erupted.

The Revolutionary Leaders

There have been reports in the press of the presence within the ghetto of armed revolutionary leaders, financed by foreign governments, and advocating communist-inspired strategies and ideologies. Our staff members have not come across leaders of this orientation. Presumably, they are the most invisible and inaccessible of the militant groups. We mention them here only to note the gap in our information. It is possible that their numbers are small and their influence so negligible that the absence of information about them is of no great importance to our understanding of the new ghetto leadership.

Militant Middle-class Leaders

A dramatic part of the phenomenon of contemporary activism is the number of traditional, civil-rights oriented, middle-class Negroes who have shifted to the militant position. Feeling that the aim of desegregation is hopeless, and that they have wasted their time in wishful thinking and futile action, they have responded with bitterness and sharply articulated hostile attitudes toward the liberal whites who were formerly their allies. They use the technics of the rebuff and the insult to shake up white people and make them realize the magnitude of the social problem facing them.

---

*Discussion of actual behavior during a riot is not a part of this paper but it should be pointed out that "having to go along with the action" describes the well-known contagiousness of crowd behavior. In the melee, one person, even a leader, could not hold out against the many, even if he wished to. But beyond the physical impossibility of individualized behavior, it becomes a matter of "face" to join one's brothers in risk-taking and a display of aggression.

That these technics usually stimulate fear and retaliatory hostility is of no concern to them. They believe that their former, progressive, non-violent alliance with whites was a mask covering the essential resistance to change dominating the whole community. It is better, they feel, to permit the real feelings of opposition to change to come to the surface than to involve themselves and their real allies—the poor, black people in the ghetto—in a game of deceit.

These middle-class leaders are composed of intellectuals and professionals: school teachers, ministers, lawyers, doctors and dentists, welfare workers and writers. Very few well-to-do Negro businessmen have as yet joined their ranks. Many of the professionals who had migrated to the suburbs of the city have moved back to the ghetto, where they are highly visible. There, they organize the self-help, community action programs described above. They try to create different pathways to the goal of stimulating, uniting, and generally lifting up the depressed mass of people in the slums. At the same time, they direct aggressive pressure, short of violence, at civic leaders, politicians, educators, and business executives, in order to get them to change their policies and practices which discriminate against Negroes. While they are convinced that only aggressive pressure by a unified black community under the control of its own leaders can produce any change, they also fear, as do most others in the ghetto, that the white community will never permit this to happen without a violent struggle.

There is one other related, behavioral phenomenon, characteristic of all the militant leadership styles which we have not yet discussed. The various community leaders, and the organizations which they represent, find it almost impossible to cooperate with each other. During a community crisis, for example, during and for a while after a riot, they may join to correlate policies. Such short-term cooperative efforts soon fall apart, and the leaders then return to their more usual competitive patterns. Intense rivalries develop not only among the different styles of leadership, but within them. The leaders vie with each other to capture a following, and, just as they attempt to discredit white leaders, so they tend to undermine each other.

The intensity of the competition has three interrelated dysfunctional effects: (1) It inhibits the unification of the ghetto which is so important to its struggles with the white power structure. (2) It makes it difficult for white, civic leaders to know with whom to deal. Who speaks for the ghetto? Which leader commands sufficient power to carry out agreed upon policies? (3) It results in rapid turnover within the leadership ranks. The leader either becomes exhausted from the in-fighting and gives up, or else his influence is eroded and he has to be replaced.

The Dynamics of Militant Activism

If these are the styles and problems of the new, militant, Negro leadership, then we are in a position to approach the question raised at the beginning of the chapter. Which factor takes the lead in defining a changing style of leadership—the requirements of the social environment or the personality of the leader? Or is the situation so much a product of interaction that neither factor can be ascribed priority? It is not the purpose of this chapter to present a definitive answer to the question but merely to contribute to a discussion of it. The strategy of discussion will consist of examining the social factors first and then see how much of the phenomenon remains to be explained by psychological factors.

The environmental factors begin with an historical situation—the collapse of the Civil Rights Movement. As long as black people both in the North and the South, perceived themselves as situated completely outside the white social system, they were eager to join with sympathetic whites to press for admission, for the "integration" and "equality" which the American system of values guaranteed. But when, due to more than a decade of vigorous civil rights agitation and legislation, they were at last able to get a foot in the door, they discovered that they were not going to be able to get the rest of the leg through, much less the whole body. At the current rate of progress, integration would take fifty to a hundred years.

With this discovery came a reassessment of the intensity of white resistance to change, of the "racism" which black people had always faced but which was supposed to be diminishing. No explanation, particularly of a social science type, based on the resistance to change of any social institution, whether black or white, could stem the tide of disillusion. "Relative deprivation" which means that a people know from direct experience what others have which they should have, indeed, have been promised, but are prevented from obtaining, and its emotional consequences, disappointment and rage, threatened to overwhelm black communities with depression or violence.

The tactical consequences of this disillusion were logical but drastic. Black people could not depend upon white people to carry out their promises. They would have to manage their own affairs as best they could. This sentiment was wide spread and spontaneous. Negro leaders did not need to stimulate it. Rather, they were required to catch up with it, to articulate it, and to formulate its expression in action. When, during Meredith's march in Mississippi in the summer of 1966, the marchers without quite realizing its import, spontaneously chanted "Black Power" among a number of other slogans invented on the spot. Stokely Carmichael merely reflected the phrase. He did not originate it, but he immediately began to give the words shape and meaning, in accordance with his personal experience, and his own style of expression.*

This sequence of events seems to represent a model for the development of new styles of militant leadership. An actual or potential leader senses the ill-formulated direction of opinion and mood among the people. Depending upon his personal tastes and background, he swims with one or another current within the general flow of opinion and action until he is able to propel himself, or is propelled, to its forefront. In the majority of cases, there is a serious question whether a particular leader actively wanted the lead, was reluctantly pushed into it by a consensus of followers seeking a leader, or was fortuitously designated a leader by a complicated process of elimination, the significance of which no one grasped at the time. It is our impression that few leaders have really understood the process through which they attained leadership, both with respect to their own motivation and that of the group.

At any rate, once at the head of a current of belief and nascent action, the leader then begins more actively to spell out future directions. Now, his own personality and motivations begin to shape an organization. Narcissistic considerations require that he differentiate himself and his followers from others. In the absence of a tradition of leadership, occasioned by the split with patterns of civil rights organizations (and middle-class styles of organization in general), the leader is free to develop his own pattern. The extent and type of innovation he imposes are limited only by his imagination and his capacity to persuade followers to support his novelties. It is this "innovative task," and the

*With its "nose for the news" sniffing the air, and in the midst of a rather dull phase of the march, the press promptly picked up the "Black Power" exchange at its source, thus notifying blacks that they had stumbled upon a slogan which would make whites sit up and take notice. The gradual crystalization of the mood of black communities all over the country around the words, "Black Power," was subsequently as much a product of the national media of communication as of the activities of black leaders.

rhetoric of instruction and persuasion required to accomplish it, that places the new leader so much in the role of "teacher" (or, as white critics might phrase it, "agitator").

Given the absence of structural constraints, the need for rapid individualization and differentiation, and the generalized adversary context vis-à-vis the white community, one might expect the new militant leader to display aggressive hostility toward rivals. It is possible that the release from formal social restraints of a certain amount of aggressivity makes it difficult to discriminate between various categories of opponents.* Furthermore, starting a new organization from scratch, or radically reorganizing an existing organization gives a leader little time to conduct diplomatic, coordinate activities with leaders of other, parallel organizations. Also, as was pointed out by Greer,[11] the maintenance of harmonious relations between similar "parapolitical" organizations is greatly facilitated by the operations of "peak associations"—that is, overarching organizations with delegated powers, such as labor federations, councils of churches, and trade associations. Due to the newness and rawness of militant Negro activism, such nationally based associations do not exist, despite the fears of those who speak of a conspiracy of violence conducted by "outside agitators." *Ad hoc* Black Power conferences and caucuses have been held to stimulate consensus among local leaders, but they are no substitute for the disciplined formulation of rules, regulations, and procedures to which local organizations agree to adhere.

These are some of the factors that can be adduced to support the primacy of social and situational determinants of leadership styles. While they bring us a long way down the road of explanation, they do not completely account for the quality and intensity of the hostility displayed by militant leaders toward whites and toward each other. Even if one takes appropriate account of the deep conflicts of interest between blacks and whites, the relative deprivation of black people, the lack of inhibition in displays of aggression among working-class people as compared with the middle-class, and the culture of violence in which

*It would appear that any political style associated with a reform movement based on radical social cleavage can give rise to belligerent rivalry and innovation, whether or not it takes place in a traditional context of roles. For example Huey Long, both as Governor and as Senator from Louisiana, displayed an inordinate amount of rudeness and uncivil behavior toward political opponents, visiting diplomats, and members of his own family and was applauded for his behavior by his followers.[10]

ghetto youth grow up, there is an edge of threat and menace, a tone of intimidation and abuse that goes beyond the open defiance and combative opposition which one would expect if social conflict and change were the only determinant of the behavior. To account for this degree of contentious, assaultive behavior, we may consider the inner, psychological dynamics of the militant position. Are there unconscious, irrational determinants mixed in with the instrumental, socially determined components of militancy? In asking this question, we do not wish to be understood as invoking any special degree of irrationality as a general explanation of Negro behavior. For example, we would suggest that the same question be asked with reference to the militancy of labor organizers during the period of the labor uprisings in the late 19th and early 20th Centuries.

We shall here simplify and limit the complex question of psychodynamics to two issues: (1) the chronic ambivalence of relations between blacks and whites in our country; and (2) the special response to powerlessness among those who are beginning to discover the uses of power.

Since the days of slavery and after, whites have had two incompatible attitudes toward blacks. The Negro is seen as slave or servant with the white man as his master—or, the Negro is seen as a compatriot who happens to be black. No matter how enlightened it becomes, the first view retains the idea of property ownership, and the Negro is in the position of an animal who may be valuable or worthless, or a servant who may be loyal or undependable. If the Negro is a good animal or servant—that is, unobtrusive and hardworking—then it is important to take good care of him. If he is a bad animal, vicious or unproductive, or a bad servant, lazy or insubordinate, then it is important to punish him or get rid of him. Seeing the Negro as a fellow citizen, on the other hand, entails no special obligation to care for him or to control him as he will do this for himself.

These two contradictory social definitions of the black man have always coexisted. They represent the ambivalent conflict of white people. If the Negro is an animal or servant, fit only for exploitation, then he must be quartered somewhere as animals are kept in kennel, stable, or field, and servants in their quarters. This is the policy of segregation. If the black man is an ordinary, fellow citizen, then he lives and works as others do—that is, in accordance with his wishes and capacities and without regard to his color. This is the principle of "integration."

"Desegregation," the traditional aim of the Civil Rights Movement, is a compromise solution of these two contradictory policies. By "desegregation," white people mean that they plan to give some Negroes a limited opportunity to be quartered amongst themselves. The "quartering" principle is preserved, but its harshness is softened. Thus the guilt which white people experience because of their ambivalence is partly neutralized. Nevertheless, this solution does not really do away with the basic incompatibility of motives. Therefore, like most makeshift compromises, it is fragile, and the basic conflict is always in danger of reappearing.*

If the "desegregation" compromise breaks down—or, if it has never been accepted—a white person usually deals with his ambivalence in a more projective fashion. White "racists" and bigots repress and project the energy of their own integrationist thinking, seeing it as emanating exclusively from their liberal adversaries, the "nigger lovers." Dedicated white liberals and civil rights advocates repress and project their segregationist ("racist") thoughts, viewing such ideas as the exclusive property of fascists and fanatics, the "nigger haters," who are easily dismissed as hate mongers by trade. Other whites, less inclined to take ideological positions on social issues, are content to display the inconsistency which their ambivalence calls out adopting integrationist attitudes in some circumstances, segregationist points of view in others.

On the part of black people, the ambivalence has been more complex and its consequences more drastic. Living as the servants of whites, doing menial work and excluded from decision-making, they have also had to learn the dangers of challenging the master race. They have experienced the "legitimate violence" of whites for minor infractions of

*The two contradictory definitions of black people arise from social and economic circumstances. The traditional economy of the rural South required the exploitation of Negro labor and therefore sought a social definition of the black man which would justify the exploitation. Insofar as the industrial economy of the North has required a large pool of skilled labor, northern metropolitan areas could not afford "racist" definitions which excluded black people (or other minority groups) from the manpower pools, and thus from the opportunity for advancement. But such social definitions are reinforced by a variety of psychological processes. Perceptions of a black skin, for example, oriented in the first instance to the socially derived, "racist" definition, soon become contaminated with primary process thinking in which "black" is equated with dirt, disorder, or uncontrolled impulses. These, and other psychological processes in whites and Negroes concerned with cross racial perceptions have been described by Pinderhughes.[12]

the rule of servility and subordination. The anger piling up within as a result of the unfairness of the system of exploitation plus the fear of showing any outward sign of displeasure toward the aggressor has been controlled by an identification with the aggressor. To the extent that the identification works, black people have seen themselves through the eyes of whites as deserving no better treatment and have turned their aggression back on themselves. Coordinate with this disposition of hostile energies, whites have been regarded with affection and admiration and their attitudes, suitably purified, have formed the basis of ego ideals. Thus black people learned not to trust each other but to depend upon white people to protect them and care for them as long—it goes without saying—as they behaved well towards whites.

There is some question about the degree to which this set of emotional controls has been conscious or repressed. Apparently, with some individual Negroes the attitude of deference towards whites and the ego control required to maintain it has been wholly unconscious. With others, there has been a consciously maintained split in ego attitudes, deference being shown only in the presence of whites. When within their own group, such individuals have always been able to make fun of the white man, or curse him out. The term, "Uncle Tom," as used by Negroes for blacks who accept white dominance, does not discriminate between conscious or unconsciously maintained controls. At any rate, whether consciously manipulated or compulsively fixed in behavior, deference based on identification with the aggressor was the chief personality pattern for black people locked into the white system of dominance.

With the popularization of the goal of desegregation during the recent upsurge in the Civil Rights Movement, this set of personality mechanisms was no longer appropriate nor necessary. The white aggressor was, presumably, disappearing. Therefore, protest, shows of displeasure, opposition to the white power structure and other examples of moderate aggression such as civil disobedience were not so dangerous. As anxiety stimulated by the environment lessened, a redistribution of internal energies was required. No longer was it so necessary to invest the image of the white aggressor with affection and admiration in order to control dangerous quantities of internal aggression which might otherwise be expressed in action. But the withdrawal of libido from the white imago led to a dilemma: What generalized object now presented itself for libidinal investment? And for identification? The white liberal was a possibility. But that depended upon what he meant by "integration" and the sincerity and effectiveness with which he could push for the establishment of more equable social arrangements. This new libidinal object was—and still is—widely accepted in fantasy among black people, but is failing to meet the test of reality. This is the point at which the disillusion, mentioned above, enters the picture. It is quite similar to the disillusion which children generally experience with reference to the idealized, fantasy parent. As it is with children, so with blacks, the disillusionment stimulated fresh anger and a realization of the necessity to do without dependence on the object whose relative weakness has now been exposed.

Thus the final solution of the dilemma, it would appear, has been directed toward the transfer of libido to the self by the incorporation into the personality of an idealized image of a black person purified of white contamination. Under such a solution, the identifications of black people—and thus, the *sense of identity*—would be wholly black. But such a solution is just now being worked out in fantasy and cannot be called an operating solution. Psychological processes laid down in childhood are not to be changed overnight. Accordingly, what is evident among militant Negroes at the present time is a wholesale display of *negative identity*. While the black identifications exist as wish-fulfilling fantasies,* in behaviour such black people are occupied in defining what they are *not*: how they are not like whites and don't like whites; how they are not like middle-class people and reject their values; how they are not of Western European origin, with several choices of alternative origins, cultures, and languages.

Negative identifications can be considered way stations on the road to a new positive identity. They are required to clear the mind of past identifications which are no longer useful, while the new idealized object is being constructed both in fantasy and in vigorous, external conflict and colloquy. In the meantime, however, they weaken the ego's control over pent up, long accumulated internal aggression. The resulting disposition to act out the aggression can be controlled only through group mechanisms which must give them some (hopefully nondestructive) outlets. This constitutes the first reason, suggested above, for the

---

*The fantasy nature of black identifications has been discussed by Pinderhughes and LeVine[13] who state, "Black people are in a state of fantasied familiarity with one another, associated with hostility toward a common enemy as they seek to establish bonds and a viable group."

particular "bite" of aggression which militant black leaders express toward each other and toward whites.

The second reason for the deep, cutting quality of militant opposition arises from the first discovery of power—or, better, the potential uses of power—among those who have been powerless. Power has many uses, not all of which are instrumental and directed toward pragmatic goals. One of the more expressive uses of power is concerned with a reversal of roles: those who have been the victims of powerlessness enjoy victimizing their supposed superiors. One sees this in small children who have discovered the use of their teeth and enjoy biting their parents—sometimes inflicting deep gashes—for the pure pleasure of knowing that they have the capacity to hurt as well as to be hurt. The knifelike edge of black militancy is a way of testing the capacity to hurt the white man and disturb his indifference or patronizing attention. Can he be frightened? Will he overrespond to intimidation? Will he express surprise, shock, anything to indicate that he now recognizes with whom he has to deal—not a thing nor a helpless child, not a savage nor a servant, not an alien nor a protegé, but a person, an individual who will take care of and defend himself in ways that are as reasonable or unreasonable as those of whites?

These consequences of Negro ambivalence toward whites are currently encountering the white ambivalence toward blacks, discussed above. It is beyond the scope of this paper to discuss the reactions of white people, except insofar as they influence black responses. But, since we are dealing with a continuous interaction, it can be observed that, at present, ambivalence weakens the capacity of many white people to understand black militancy. They continue to overrespond to intimidation as if they expect black people to fulfil their fantasy of revenge: Having been subjugated and having discovered power, they will now try to subjugate white people—a goal to which no black leader has subscribed. They misunderstand black threats, taking them as final intentions of revenge rather than tests of strength. They reason that if black people will not subscribe to their version of "desegregation"—the cure for the persistent gnaw of white guilt—then they must have revenge in mind. Accordingly, they talk of protecting themselves from black violence, not perceiving how black people will respond to this fresh demonstration of blindness to the real issues of freedom. On their side, black people misapprehend the psychological basis of white defensiveness and expect white people to institute a new and more violent repression.

In doing psychotherapy, we assume that neurotic symptoms are the result of compromise solutions of unconscious conflicts based on ambivalent feelings. As long as unconscious defenses keep the ambivalence of feeling—and its sources in memory—outside of awareness, no better solutions can be worked out. With the weakening of defenses, the feelings plus their historical roots can enter consciousness and be modified. More adaptive compromises can be tried out in experience.

We possess no parallel psychotherapy for social processes such as race relations. Since there is no "sociotherapy," the exposure of the unconscious ambivalence and unworkable compromises underlying racial tensions can emerge only from overt social conflict which forces feelings into awareness by a kind of "shock therapy." Clumsy though it is, entailing more hardship than necessary, the method will work, and better solutions will ultimately be discovered. But those whites who are involved in understanding human behavior, the psychiatrists, social scientists and philosophers active in the intellectual establishment, can speed the process of discovery. What is required is effort and self scrutiny. Because black militant leadership is currently the leading edge of innovation and discovery, every effort should be made by white intellectuals to understand it and themselves and to help bring white responses into adaptive contact with it. To be adaptive means not to avoid conflict when it is based in reality. Only by recognizing this fact can white intellectuals contribute to the solution of the race problem.

REFERENCES

1. Lasswell, Harold D.: Power and Personality. New York, W. W. Norton and Co., 1948. Also: Psychopathology and Politics. New York, Viking Press, 1960.
2. Lane, Robert E.: Political Personality and Electoral Choice. American Political Science Review, XLIX, 173-190, 1955.
3. Downs, Anthony: An Economic Theory of Democracy. New York, Harper and Brothers, 1957.
4. Mussen, Paul H. and Wyszinski, Anne B.: Personality and Political Participation. Human Relations, V, 65-82, 1952.
5. DiRenzo, Gordon J.: Professional Politicians and Personality Structure. Am. J. Socio., 73, 217-225, 1967.
6. Adorno, T. J., Frenkel-Bruswik, Else, Levinson, D. J., and Sanford, R. N.: The Authoritarian Personality. New York, Harper and Brothers, 1950.
7. Dicks, Henry V.: Personality Traits and National Socialist Ideology. Human

Relations, III, 111-153, 1950.
8. Carmichael, Stokely and Hamilton, Charles V.: Black Power: The Politics of Liberation in America. New York, Random House (Vintage Books), 1967, p. 15.
9. Powledge, Fred: Black Power—White Resistance: Notes on the New Civil War. Cleveland, World Publishing Co., 1967.
10. Hess, Stephen: America's Political Dynasties: From Adams to Kennedy. New York, Doubleday, 1966, Chap. 15, "The Long Dynasty."
11. Greer, Scott and Orleans, Peter: Political Sociology. *In* Robert E. L. Faris (Ed.): Handbook of Modern Sociology. Chicago, Rand McNally, 1964.
12. Pinderhughes, Charles A.: The Psychodynamics of Dissent: A Clinical Appraisal with Emphasis on Racial Activists. This volume, Part II, Chap. 4.
13. Pinderhughes, Charles A. and LeVine, Herbert O.: The Psychology of Adolescents in a Peaceful Protest and in an Urban Riot. Paper delivered at the Divisional Meeting of the American Psychiatric Association, New York, November 6, 1967.

Discussion by Brahm Baittle, M.D.

Dr. Spiegel has taken pains to demonstrate that there is not one type of activist, but many. He suggests they can be grouped into five phenomenologically different categories, each with distinctive dynamics. Although each of these groups may be somewhat homogeneous by these criteria, Dr. Spiegel implies that different character types are to be found within each group. However, in the last section of his paper, Dr. Spiegel formulates dynamic constellations that he feels are common to all activist leaders, a type of generalization that is unwarranted in my opinion, especially since he himself stressed the limitations of the data collected. It seems to me also that the historical and social stresses he emphasizes affect most lower class Negroes, and he does not attempt to explain in what way and for what reasons the activist leaders respond differently than the rest.

Dr. Spiegel makes very limited generalizations, confining himself to those attributes common to all activist leaders on the basis of which they can organize.[1] An alternative hypothesis is suggested by Coles,[2] who argues in *Children of Crises* that the only characteristic in common is an "ethical" motivation in a moment of opportunity in history. This may become attached to any character trait and become interwoven into the character structure. Thus Coles claims that we cannot generalize activists according to character types. The character traits are nonspecific and act as levers for the more conscious and preconscious ethical and idealistic motivations. Of course, Coles was describing for the most part nonviolent southern activists, but his hypothesis could be considered as a reasonable alternative.

Dr. Spiegel describes two dynamic constellations that seem common to activist leaders. One set centers on the ambivalence of Negroes to whites, with subsequent idealization and dependence on whites, identification with the aggressor, and identity problems. The other concerns "the special response to powerlessness among those who are beginning to discover the uses of power". From a psychoanalytic perspective, these formulations are mainly from the adaptive point of view, and to a lesser extent from the dynamic point of view. This is the characteristic stance of most social psychological research with an analytic orientation. However, this is not the only, or even main, contribution that psychoanalysis can make in studies such as these. One of the most significant attributes of psychoanalysis is that it is a genetic psychology. The concept of psychic structure is a relevant feature. The quantitative or economic factor is yet another viewpoint from which data can be analyzed.

Care must be taken lest concepts of one discipline are applied automatically to concepts of another, thus falling into the trap of reductionism. The terms used by Dr. Spiegel in the psychological parts of this paper are psychoanalytic and refer to intrapsychic processes. There may be analogous concepts in interpersonal and sociological theory, and bridging concepts are needed to formulate the relationships between the two. Unfortunately, in this chapter, he has not kept the disciplines separate and has confused intrapsychic, sociological, and historical frames of reference. For example, in discussing one side of the militant Negro leader's ambivalence to the whites, Dr. Spiegel formulates the dynamics of the idealized and admired object. However, he fails to probe into the genetic, structural, and economic implications of his dynamic formulations from an intrapsychic point of view, and thus is unable to correlate these formulations nor speculate on bridges to the social dynamics that he does describe. The continuing discussion of this chapter will offer clues, I hope, to areas where articulation between the psychoanalytic and sociological concepts may take place.

Dr. Spiegel's formulations about both the idealized object and the object as aggressor lead to further questions. Does Dr. Spiegel feel these are external objects or internal? If the latter, it would be of interest to determine their location in the topographic map. Their relationship to drives and the superego would be important to work out. Little attention has been paid to the internalization process after the oedipal period. Perhaps studies such as this one may throw light on such processes because the internalization of the white man by the Negro, and vice versa, is of such importance and visibility. This also brings up the question as to what extent these later internalizations are displacements of parental images.

If these objects are internalized, it is pertinent to ask when it occurs. Dr. Spiegel talks of a "collective narcissistic injury"[1] to account for this type of object relation. I believe this concept needs further elucidation. When does this narcissistic injury take place? The answer to this question will have structural and economic implications. A hypothesis that an external chronic frustration of recent historical origin leading to a sort of traumatic neurosis in a psychic structure already mature and more or less well functioning could be formulated. An alternative hypothesis may hold that the injury occurs in childhood, for example in the second or third year in life when the parents are assigned the omnipotence that was once the child's. This alternate hypothesis would imply that the parents are the mediators of the culture and a study of child rearing practices would be in order. To what extent are child rearing practices influenced, say, by an isolated Negro culture as one variable, and the culture of the relationship between the Negro parents and the whites as another variable? Other stages of child development, especially adolescence, could be hypothesized as alternate or additional vulnerable periods to this type of stress. Depending on the quality, intensity and duration of the narcissistic injury, the age at which it occurs, past experiences, and the current organization of psychic structure, the child or adult will cope with the stress in many different ways. Data such as these are hard to collect and harder to assess, and could not be expected from the methods of collecting data in this study so far. Yet these questions do help to spell out the limitations more precisely that Dr. Spiegel mentioned. One would hope in future studies that more intensive interviews could be carried out, if this will prove to be feasible.

Another suggestion for future investigation would be the study of a control group of activists who were not leaders. In addition to acting as a control, studies could be made on the influence of group formation itself on the behavior of the activists. The dynamics of some groups have been studied under stress[3], and it would be informative to compare and contrast these to the dynamics of group phenomena under study in this project.

Similar types of questions arise over the formulations about identity. The process of achieving identity is a long and complicated one, involving phenomena that need to be conceptualized both psychologically and sociologically. Erikson[4] points out that identity begins with the achievement of basic trust in the earliest years of life. The differentiation between self and object, object constancy, the achievement of autonomy in the anal period, sexual differentiation and a myriad of other factors throughout development contribute to a sense of identity. Along with intrapsychic growth of a sense of identity is the need for recognition by the outside world of that identity. The parent's contribution is the only significant one in the first years of life, but when late adolescence and early adulthood is reached, society's recognition is critical. Dr. Spiegel has described the negative identity of the Negro activist leader. This leads to the further question of the locus of the trauma. Does it lie mainly in the failure of the white society to acknowledge a positive healthy Negro identity? If so, the age when this happens is crucial. Alternatively, this negative identity may have its roots in infantile intrapsychic identity problems. Most likely, combinations of both differ in various people. One may ask if people with preexisting identity pathology are more apt to establish the negative identity that Dr. Spiegel describes. I do not think there is an obvious answer. Further study is needed to elucidate this and many other problems that Dr. Spiegel has brought to our attention in this preliminary review.

References

1. Spiegel, J. P.: Personal communication, 1967.
2. Coles, R.: Children of Crisis. A Study of Courage and Fear. Boston, Toronto. Little Brown and Company, 1967.
3. Janis, I. L.: Group Identification Under Conditions of External Danger. Brit. J. Med. Psychol. 36:227-238, 1963.
4. Erikson, E. H.: Identity and the Life Cycle. Psychological Issues 1:101-164, 1959.

THE FUNCTIONS OF DISSENT

By LEWIS A. COSER, Ph.D.

MOST OF THE FREEDOMS which we hold dear have been won by rebels and radicals. This fact creates a curious paradox for those social scientists who have of late become enamoured with system theory, or at least its more conservative variants: The most desirable features of the present system of wont and use have in fact been the result of breaches of previous systems of wont and use. The system as it is today emerged to a significant extent from successful efforts of dissenters and radicals to attack some of the very foundations of antecedent systems. This is why the rebels of yesteryear are among our most venerated culture heroes and exemplars.

The apparent paradox that seems to be involved here may be resolved if we discard the common-sensical view which has it that any deviance from the norms involves a weakening of social systems, and consider instead the notion that deviance may, under specifiable conditions, help strengthen the solidarity and cohesion or restructure the group in which it occurs.

Even criminal deviance from the norms of a society, far from necessarily having negative consequences, may have a series of functional consequences for that system.[1] This becomes readily apparent if we discard the tendency, which is natural for a psychiatrist, to probe into the motivations of the criminal deviant and instead redirect attention to the unanticipated societal consequences of purposive criminal behavior. Once the perspective is adopted it can be shown that behavior that involves a motivated defiance of group norms may yet in its social consequences lead to a strengthening of the norms, the cohesion and the morale of a group or of a total society. Karl Marx, George Herbert Mead and Emile Durkheim have all drawn our attention to the fact that the very infringement of the norms of a group may lead to the reactivation of previously inactive defense mechanisms. "Crime," wrote Durkheim, "brings together upright consciences and concentrates them." And Mead argued in a similar vein: "The criminal . . . is responsible for a sense of solidarity, aroused among those whose attitudes would otherwise be centered upon interests quite divergent from each other."

Following the Durkheimian tradition a modern sociologist, Kai T. Erikson, argues that any system must have mechanisms that maintain boundaries which mark it off from the surrounding environment, and then goes on to state that, ". . . members of a community inform one another about the placement of their boundaries by participating in the confrontations which occur when persons who venture out to the edges of the group are met by policing agents whose special business it is to guard the cultural integrity of the community."[2] The definition of what is considered normal in the group takes place with reference to what is considered deviant, and normal morality is given its content through the contrast provided by that which is not normal. To use the language of Gestalt psychology for a moment: just as figures cannot be perceived except in relation to grounds setting them off, so normality can hardly be perceived except in contrast to deviance; to be "good" makes sense only in relation to being 'bad." This is, of course, what Hegel meant when he argued in defense of his dialectical method that the positive and the negative gain their "truth only in their relation to each other so that each contains the other within it."

If we discard the common-sense idea that criminal deviance necessarily redounds to the detriment of the group in which it occurs, and accept instead the notion that the very attack against the group's norms which criminal deviance implies may in fact strengthen it through the arousal of defense mechanisms, then the notion that rebels and dissenters serve positive functions becomes a fortiori more plausible.

Dissenters and radicals, rather than being motivated like the criminal, by private desires for the maximization of self interests even at the price of normative deviance, are in fact motivated by the desire to improve the society's conditions. That is, while we notice a disjunction between private motives and public consequences when we consider criminal behavior, such a disjunction is not characteristic of the innovating behavior of the dissenter and radical, even though it may, of course, be present in many individual cases. Only a profoundly reactionary type of reasoning assumes that dissent is explainable in terms of the private sickness in the dissenter's soul. We must assume that such dissent may be more readily explained as a reaction to what is perceived as a sickness in the body social. There are, of course, many dissenters who are sick; but it is an open empirical question whether sickness is more prevalent among them than among the stout upholders of conventional morality.

In certain cases useful analytical purposes may be served by classifying all behavior which deviates from the norms under one conceptual category of deviance, yet it must also be kept in mind that the varieties of deviant behavior differ in their structural, cultural, and functional consequences. To be sure, the dissenter and the criminal defy normative expectations, yet they are profoundly dissimilar also; nonconformist dissent, "is not a private dereliction, but a thrust toward a new morality or a restoration of an old and almost forgotten morality . . ."[3] Dissent, as distinct from crime, involves the striving toward an alternative moral basis rather than simple moral deviation.

The distinction involved here may be highlighted when we reflect that the criminal typically operates in secrecy, and tries to minimize the chances of detection. The dissenter in contrast operates, so to speak, in broad daylight—at least in open societies—in order to attract a maximum audience. The criminal wishes to shield his operations from observability by others, the dissenter welcomes all chances to maximize the visibility of his actions. Being oriented toward the collectivity, he is led to seek and find an audience within it, and he strives to attain maximum publicity for his message.

Robert K. Merton has specified a number of other dimensions in which criminal deviance can be distinguished from innovating dissent and I have contributed some supplementary considerations to his seminal ideas.[4] For our purposes it seems sufficient to note one additional major distinction.

Criminal deviance, though it may upon occasion lead to change in, say, the legal or the police system, has yet essentially conservative consequences for the society. It gives an impetus to the conforming members of the society to close rank and to shore up the threatened stability of the system. Dissent, on the other hand, is not only consciously designed to bring about change and innovation, it also tends, when the circumstances are propitious, to institute such change.

The innovations that the dissenter proposes to the group may be received with variant degrees of tolerance, depending on the character and values of the group. In certain settings, say in scientific societies, innovating proposals may in fact be highly valued. Here we deal with a special type of conformity rather than with deviation.

Total societies, however, are never likely to provide so propitious a setting for innovation. Vested interests in the prevailing system of wont and use are always likely to perceive the innovator as a threat to the system and hence to resist him.

In relatively static societies, the tradition of the eternal yesterday, to use Max Weber's phrase, lies so heavily upon the land that dissent tends to be perceived as an attack upon the total system and to be rejected out of hand. In such societies the chances are indeed high that the dissenter will be defined as a heretic and cast out or destroyed.

Matters are however quite different when we consider rapidly changing societies such as those that have existed in the West for at least two hundred years. Here the rapidity of change in various institutional areas, from technology to the polity, has created a much more fluid situation in which receptivity to dissenting opinion has been increased. Such societies are characterized, to mention just one of their distinctive attributes, by high degrees of heterogeneity. They are likely to present a picture of clashing and contending subgroups and subcultures. Moreover, unequal rates of change are observable in different institutional orders of such societies. While change may be rapid, say, in the sphere of technology, the legal system may lag behind and may not have yet caught up with revolutionary developments in the technological sphere. The failure of family law to come to terms with the recent revolution in biological engineering, e.g., artificial insemination, may serve as a convenient example. In similar ways, the moral order may only imperfectly adjust itself to rapid changes and innovations in the technological scene. Witness the as yet most halting adjustment of sexual mores to the availability of new contraceptive devices which now almost completely permit sexual enjoyment divorced from reproduction. Examples could be multiplied.

We need not limit ourselves to the impact of technological changes. Consider only the fact that in recent decades America has been experiencing a rapid acceleration of trends to evaluate its citizens by universalistic rather than by particularistic criteria, by what they do rather than by who they are. The ethos of achievement, of course always very strong in American society, has been stressed in the more recent decades with unequaled emphasis. Yet we notice in a variety of areas an enormous reluctance to admit in practice what might be affirmed in theory. Thus, even though the law has been readjusted, the mores still put up an enormous resistance to real equality, that is real universalism, in the treatment of the American Negro. Be it in education or in housing or in employment, the walls that shut out the Negro have by no means come down, in fact, as in urban housing in the North, they have probably grown higher. The discrepancy between doctrinal affirmation and actual behavior in this sphere is enormous.

In rapidly changing societies, then, the old certainties are no longer unshaken and traditional wisdom is no longer sufficient to cope with the challenge of the day. In such societies, it stands to reason, innovating and dissenting voices are likely to get a measure of hearing. At least some of the component individuals and subgroups in such a society will perceive the dissenter and innovator as intending to perform positive tasks, even though they may initially judge his activities to be misguided.

One of the major functions of the dissenter, it would seem, is that he provides an angle of vision which allows him to perceive certain flaws in the existing order of things that remain hidden from the upholders of tradition. The sociology of social perception still has to be written. We still do not really know why people suddenly see certain blights which previously remained unperceived.[5] Why, for example, did solid middle-class America, as well as the powers that be, virtually ignore the problem of poverty in the midst of plenty throughout the fifties? And why has poverty loomed so very large in the preoccupations of the sixties? No ready answer can be given, I believe, but one incontrovertible fact stands out. Had it not been for a small band of dissenters, from Michael Harrington to Dwight Macdonald, it is most unlikely that the war against poverty would ever have been inaugurated at this time. It is only when a variety of pamphlets and books from such dissenting voices began to assault middle-class consciousness, and when the mass media began to distribute these messages, that the powers were prodded into action. There are historical and structural reasons to account for the new receptivity of the public and for its willingness now to confront what it previously failed to perceive. But the most elaborate stage setting will not make for a play, if actors fail to make an appearance on the scene. Michael Harrington, to be sure, did not create the war on poverty, and Martin Luther King did not create the Civil Rights Movement, but without their appearance preexisting scenarios would not have been incorporated into the resistant texture of American reality.

It would be superficial, however, to restrict one's focus to those instances in which innovation succeeds by stirring a receptive audience into action. We must also consider that heterogeneous societies, such as ours, include in their midst individuals and groups that are highly motivated to resist innovation with furious energy. Richard Hofstadter has called this the "pseudo-conservative revolt" in recent American history.[6] Though McCarthyism and its successors on the radical right fused a number of themes in their propaganda of hate, one major theme seems to loom largest: The Revolt against Modernism. One senses here a complex feeling of revulsion against the very idea of modernity. Many of the fundamentalists who are susceptible to such appeals feel cut off from their accustomed moorings and have a sense that the drift of history has passed them by. These strata fear that the vast apparatus of moral restraints embodied in the Protestant Ethic has been cast aside in the modern world. They have come to feel that the virtues of prudence, economy and diligence have been flouted and that the very ascetic disciplines by which they governed their lives have been cast aside. Whether they think that the Whore of Babylon now sits in Moscow or in Washington seems less important than that they respond to a paranoid sense of being surrounded by a vast conspiracy of men eager to undercut the very assumptions and premises upon which they have built their lives. Their desperate resistance to change can hardly be overestimated. They fight for their lives.

Whether vested interests resisting change or groups and individuals with a more flexible stance will dominate in a society will depend on the constellation of specific historical forces as well as on the character of existing social structures. Flexible social systems and well integrated social structures are more likely to welcome dissent and conflicts which might lead to genuine transvaluations, while rigid structures tend to repress them.

But beyond such general characteristics of historical and social constellations, we must also consider the particularities of specific movements of dissent and protest. It is by no means the case that all such movements are likely to have the same repercussions in the public at large. Movements, for example, that are based on the implementation of previous value commitments of a society are likely to be given a different reception than movements that propose a fundamental change in these commitments. Kenneth Boulding provides an instructive example in a discussion of the protest against the war in Vietnam and the civil rights protest.[7] In the latter case it was very clear that "certain fundamental images of the American society were inconsistent with its practices, and therefore, the protesters could appeal to an ideal which was very widely held. Wherever there is hypocrisy, there is strong hope of change, for the hypocrite is terribly vulnerable to protest." The civil rights issue was a case in which the society was already prepared for change even though it would have been long postponed had the protest and dissent not brought it forcibly into the forefront of the nation's conscience.

Matters are, however, very different in the case of the war in Vietnam. Here the protest movement has a much harder time since it has to move society into an entirely new position rather than simply crystallizing change in a society which is fundamentally ready for it. Where a society is divided and ambivalent, a protest movement designed to push it into one direction is likely to bring forth a countermovement. Simple protest movements are hence much less likely to show positive results in the short run. Where we do not deal with a movement that seeks the crystallization of change for which a society is substantially ready, quite different technics, Boulding suggests, must be used. In the one case shrill protest using all sorts of obstreperous means to destroy the pattern of existing legitimacy in the interest of a new pattern which is waiting to be born would seem quite effective. In the other, however, an educational movement using a low-keyed strategy and respectful of existing legitimacies may be far more useful.

At an earlier stage of American race relations, the low-keyed educational efforts or organizations such as the NAACP seemed to be the most effective means for bringing about change; at a later stage, when the educational groundwork was done, the more drastic methods of the Reverend King became more effective. And at a still later stage it was perhaps inevitable that the more militant SNCC moved into the forefront.

But this hardheaded appraisal of the chances of dissenting and protest movements in different periods and at different stages is not likely to be found congenial by many dissenters. A great number of them are men in a hurry who are incapable or unwilling to postpone the satisfaction of being effective and to trust the often inevitably slow march of historical events. When they encounter the fierce resistance of vested interests and the upholders of the traditional order and when they perhaps realize that they preach only to the already converted, they may lose their initial high hopes. In such cases two pathways, both, I believe, equally ineffectual, readily suggest themselves. The dissenter may withdraw into the comparative security of a sect of like-minded "true believers," or he may attempt to join the forces of the establishment in order to bring about practical change from within.

In both cases, I believe, the dissenter is in fact neutralized. Within a sect of "true believers," he is likely to nurture apocalyptic notions which remove him from a hold on the realities of the present. His vision of the world will be shaped by sustained contact with the small saving remnant of the "faithful," and he will progressively remove himself from contact with the surrounding world. Immersed in the esoteric world of the sect, he will disdain and fail to comprehend the exoteric outside. His every action will betray his contempt for existing society—and this society will repay him by treating him with disdain or indifference. By having lost contact with society, he has ceased to be dangerous or even to claim serious attention. And, worst of all, whenever he will get attention at all, it will be as a source of amusement when he is put on exhibit in the mass media for the entertainment of the squares.

The dissenter can have an impact upon a society only to the extent that he maintains a link with that society. If he removes himself too far from its assumptions, if he moves too far out onto its margin, he no longer impinges upon it and he can be neutralized. Only, in other words, when the dissenter maintains a creative tension between himself and the society that he criticizes can he have a measure of effectiveness.

The other danger that besets the despairing dissenter faces him when the society rather than casting him out, welcomes him and hence renders him ineffective through cooptation. David Riesman some time ago coined a memorable phrase when characterizing the role of intellectuals in contemporary American society: "Were not intellectuals of more use to this country when they had less use for it?"[8] I take Mr. Riesman to mean that only those intellectuals who maintain a certain distance from their society can be effectively critical of it. Only through critical intelligence made possible by some remoteness from everyday affairs, and cultivated concern with ultimate rather than with proximate values, can they serve fully in their innovating capacities. If intellectual spokesmen get coopted into an "intellectual establishment," if they become part of a kind of clerisy, they will no longer be able to perform their distinctive critical roles. To the extent that disabused dissenters, having become wearied of the role of prophets in the wilderness, turn toward the cultural and political establishment, their voices will inevitably be stilled. Whether wittingly or unwittingly, they will slowly be transformed into experts who, instead of influencing policies, will be condemned to execute them. They may move into these positions from the purest of motive, feeling that the time has come to put the hand on the tiller, to move from a sterile criticism to responsible implementation. Yet I dare say, in most cases the house of power will so envelop and absorb them that the erstwhile dissenter will be transformed in due time into a New Samurai.

What I have been saying is that both extreme removal to the outer edges of a society and incorporation into its core are likely to mute the voices of dissent and to neutralize the dissenter. It follows that a certain degree of measured estrangement is the very precondition for playing an innovating role. Yet this detachment must at the same time be based on a deeply felt commitment to the ideals and central values on which the society rests. We are likely to be especially critical of the things we love, Those with whom we are most deeply involved, our children, for example, will be most sharply criticized precisely because we expect so very much from them. Similarly, effective dissenters may be highly critical of their society *because* they are intensely devoted to its underlying aspirations. They will criticize its current performances in the light of these aspirations. They will attack what *is* in the name of an *ought* derived from the very value premises to which the society professes to adhere. In this sense, dissenters may indeed be detached at the same time as they are deeply concerned. Such an attitude permits them, I believe, to transcend both affective neutrality and blind involvement. Through *detached concern* with the fate of their society and their fellow men avoiding the twin temptations of total withdrawal and total integration, dissenters may play a major role in modern societies.

Most observers would tend to agree that the most salient characteristic of modern societies is to be found in the highly bureaucratized and rationalized structures which have come to dominate the lives of modern man. Deliberate control and rationalized regulation of "spontaneous" processes has become the rule in political and military, in economic and scientific institutions. Bureaucratic forms of organization with their emphasis on calculable, methodical and disciplined behavior have made possible the towering achievements that characterize the history of modern man. But the very structures which men have created in their quest for dominating nature and for constructing a more rational human world increasingly have come to be viewed as Frankenstein monsters which threaten to enslave their own creators. The pervasive discontents and malaise which pervade the present cultural atmosphere may be understood, at least, in part, as being caused by our deep-seated ambivalence about the bureaucratic structures of our own creation.

What seems so particularly frightening about modern rationalized systems is that in their orientation toward ever increasing predictability and calculability of results they tend to repress spontaneity and creativity. Though they do not reject all novelty, they tend to rely only on programmed innovation and develop a built-in resistance to the challenge of the spontaneously new. Bureaucratic structures, in other words, are subject to the blight of ritualism.

Such structures are likely to rely on functional rationality, i.e., the coordination of action with reference to the goals of organizations and to distrust substantive rationality, i.e. intelligent insight into the interrelations of events.[9] They are likely, if unchallenged, increasingly to rely on habit and the routine responses to predictable stimuli. Yet, wherever habit dominates, creative thought is bound to wither away. As John Dewey once put it, "Only a hitch in the working of habit occasions emotion and provokes thought." Organizations that depend on habitual reactions are bound to ossify and to be unable to respond to the challenge of genuine novelty.

If the above description is even approximately correct, dissent can serve highly positive functions in modern bureaucratic societies. The dissenter, by challenging wont and use, provides challenges to such systems and hence helps prevent them from becoming cold, vicious and increasingly incapable of responding to new experiences and new opportunities. They need to be constantly stirred into renewed awareness if they are not to fall victim to the debilitating pull of the habitual and the appeals of precedence and tradition. If substantial rationality is to prevail over functional rationality, sluggish bureaucratic life must constantly be prodded through voices of dissent. Without the tonic virtues of dissent, modern bureaucratic societies are likely to go the way of ancient Egypt and ancient China.

REFERENCES

1. See my recent Continuities in the Study of Social Conflict. New York, The Free Press, 1967, esp. pp. 111ff. A number of ideas adumbrated in the present paper are more fully developed in this volume.
2. Erikson, Kai T.: Wayward Puritans, New York, John Wiley, 1966, p. 11.
3. Merton, Robert K.: Social Theory and Social Structure. New York, The Free Press, 1957, p. 363 *et passim.*
4. *Ibid.*, pp. 357-368, and Coser *op. cit.* 123 ff.
5. Cf. Coser, Lewis A.: The Visibility of Evil The Journal of Social Issues (forthcoming).
6. Hofstadter, Richard: The Paranoid Style in American Politics. New York, Knopf. 1965.
7. Boulding, Kenneth: What the Teach-Ins Taught Us. Dissent, Jan.-Feb. 1966, pp. 10-15.
8. Riesman, David: Comments on American Intellectuals. Daedalus, Summer 1959, p. 493. For a further elaboration of my views on American intellectuals today see my Men of Ideas. New York. The Free Press, 1965, especially the concluding chapters from which I have borrowed some of the ideas developed here.
9. Cf. Mannheim, Karl: Man and Society. *In* an Age of Reconstruction. London, Routledge and Kegan Paul, 1940, pp. 51-60.

Discussion by Milton Mazer, M.D.

Professor Coser has presented us with a closely reasoned paper on the relationship between the structure of society and social dissent. He argues that far from being destructive of social cohesion, dissent, whether directed against society as in criminal activity or for social purposes, incites defensive measures which strengthen that society. He tells us that static societies resist social dissent and that rapidly changing societies are receptive to it. He asks one of the big questions: What are the factors that permit some of us at certain times to become aroused by conditions which were previously unnoticed? Why was the fact of poverty in America so long ignored? What occurred in the last few years to make its indecency suddenly visible?

Further, he suggests two strategies for dissenters. If a subject of dissent is consistent with the professed principles of the society, then an educational movement with a low-keyed strategy will be most effective. If, on the other hand, the society has not yet even in principle recognized the problem, then the protest may have to be shrill and obstreperous in order to destroy the legitimacy of the social practice. He goes on to describe the twin dangers of the frustration of dissent. The dissenter, when frustrated, may withdraw into the security of a sect and become a "true believer," or he may join the establishment in order to bring change from within. In both cases, Professor Coser contends, his dissent is neutralized. He warns intellectuals against accepting invitations to put their ideas into practice lest they "be transformed into experts who instead of influencing policies, will be condemned to execute them". He is afraid, I gather, that the sharp edge of their dissent will be dulled by compromise, and they will end up as docile members of the establishment. Coser plumps for the middle position, a degree of "measured estrangement" or as he puts it, "the stance of the effective dissenter must be that of detached concern".

Having described Coser's view of dissent, I should like to outline another position as a basis for our discussion. In contrast to Coser who describes dissent as an intellectual process, I would suggest that there are many modes. Those who express their views by marching are, as Lenin put it about the Czar's deserting soldiers, voting with their feet; those Negroes who sat in at lunch counters in the South dissented physically just as clearly as Dwight MacDonald dissents intellectually from the offices of the New Yorker Magazine.

Second, I am suspicious, as I am of all dualisms, of Coser's separation of intellect and action, as when he warns the intellectual against accepting positions in which he has a chance to put his desires into practice. His distinction between *influencing* policies which he sees as good and *executing* policies which he sees as dangerous is another deceptive dualism. If the dissenter really wants the thing to happen, in my view he must welcome responsibility and see it through from idea to action. For wrongs are not righted by declarations and manifestos alone; between the idea and the action there are always human agents.

I believe with Holmes and with Dewey that ideas are incitements to action and that the point of dissent is not to help the dissenter remain a critic on the side of the angels. The point of dissent, is to change the world. The stance he describes for the dissenter in terms such as "measured estrangement" or "detached concern" seems to me to be descriptions of self-immobilizing caution. I am afraid that the intellectual who remains in a state of "measured estrangement" or "detached concern" will have kept his passions so pallid that he will find it more important to keep his shield polished than to get into the arena where battles are fought and institutions changed.

It is true, as Coser points out, that whenever one becomes part of an organization whether it be government, a psychoanalytic institute, or a university, that there is danger of personal subversion, of being taken over by the establishment. Along with this risk, however, goes the rare chance of putting one's purposes into effect. When Jerome Weisner became the President's Science Advisor, he ran the danger of becoming co-opted into the establishment; as it turned out, however, he had a large share in achieving the Test Ban Treaty. To avoid the risk to his own purity of purpose should he have left the task to lesser men?

I, also, find myself uneasy with Coser's venture into strategy. He defines two separate strategies for two separate social circumstances. If he is correct, then the proper strategy for the Civil Rights Movement should have been a low-keyed, educational approach, since its goals had long been part of our pretentions and had long been written into our revered documents. Had this prescription been followed Martin Luther King would not have led the bus boycott in Montgomery, the Freedom Riders would have remained at home, and the Negro children of Little Rock would not have braved the mob to get to school. Can we believe we should be farther along without these acts? I think the difficulty with Coser's strategic prescriptions is that there is an air of determinism about them. Societies, and even more importantly, individual men do not always behave as behavioral scientists predict they will. There is usually someone around who either gums up the works or greases the skids, and so confounds our predictions. Societies are not great machines which act in accordance with nature's immutable laws; human purpose, human will, human passion, do make a difference. Those engaged in dissent on the great issues of a society do not act on the basis of a strategic logic, but out of the imperative of passion. Martin Luther was not inhibited by the fact that there had already been two dozen futile attempts at a Reformation before him and that the bones of heretics lay everywhere about him in unmarked graves. Perhaps those of us who stop to plan strategy, to study the odds, are not those who change societies. Had Lenin listened to the warnings of a social scientist of his day, Karl Marx, that the Socialist Revolution could occur only be started on a soil nourished by a class conscious industrial proletariat, he might have given up the idea and never left Switzerland for the Finland Station. I think we would do well to remember Fichte's comment that "The I is not a fact but an act".

In summary, my main disagreement with Professor Coser's position is based upon what I consider a fatal separation of mind and action in his paper. I think he counts too much on the pretensions of social theory and too little on individual will. After all, social theory is always based upon the data of the past, how men have behaved, and unlike the more comfortable position of physical theory, it has no assurance that men will continue to act as they have. This is both its fate and our hope. In speaking of our legal system Justice Holmes once remarked that a law is simply a prediction of what a court will decide. I think we may say with equal validity that our social theories are, in truth or in error, no more than chancey predictions as to how men will act. And that we, who choose to act or not, may by our passivity or our passion, confound the most perceptive predictions of our theories, and even more important, we may in some measure, change the world. This, I believe, is the function of dissent.

NAME INDEX*

(Principal References in Italics)

*Prepared with the aid of Miss Savilla Laird.

SUBJECT INDEX

(Principal References in Italics)